MARINE GEOLOGICAL SURVEYING AND SAMPLING

Marine Geological Surveying and Sampling

Edited by

E. A. HAILWOOD

Oceanography Department, University of Southampton, Southampton, U.K.

and

R. B. KIDD

Geology Department, University College of Wales, Cardiff, U.K.

Reprinted from Marine Geophysical Researches
Vol. 12, Nos. 1–2 (1990)

KLUWER ACADEMIC PUBLISHERS
DORDRECHT / BOSTON / LONDON

Library of Congress Cataloging-in-Publication Data

```
Marine geological surveying and sampling / edited by Ernest A.
  Hailwood and Robert B. Kidd.
      p.   cm.
   "Papers ... from a meeting organized in May 1988 at the Geological
Society, London, under the auspices of its Marine Studies Group"-
-Introd.
   "Reprinted from Marine geophysical researches, v. 12, no. 1-2
(1990)."

   1. Submarine geology--Congresses.   2. Geological surveys-
-Congresses.  3. Sampling--Congresses.   I. Hailwood, E. A.
II. Kidd, Robert B.   III. Marine geophysical researches.
QE39.M294  1990
551.46'08--dc20                                          90-4537
```

ISBN-13: 978-94-010-6763-8   e-ISBN-13: 978-94-009-0615-0
DOI: 10.1007/978-94-009-0615-0

Published by Kluwer Academic Publishers,
P.O. Box 17, 3300 AA Dordrecht, The Netherlands.

Kluwer Academic Publishers incorporates the publishing programmes of
Martinus Nijhoff, Dr W. Junk, D. Reidel, and MTP Press.

Sold and distributed in the U.S.A. and Canada
by Kluwer Academic Publishers,
101 Philip Drive, Norwell, MA 02061, U.S.A.

In all other countries, sold and distributed
by Kluwer Academic Publishers Group,
P.O. Box 322, 3300 AH Dordrecht, The Netherlands.

printed on acid-free paper

Contents

E. A. HAILWOOD and R. B. KIDD / Editorial Introduction   vii

PART I: SURVEYING TECHNIQUES

K. G. ROBERTSON / Deep Sea Navigation Techniques   3

J. A. GRANT and R. SCHREIBER / Modern Swathe Sounding and Sub-Bottom Profiling Technology for Research Applications: The Atlas Hydrosweep and Parasound Systems   9

R. C. SEARLE, T. P. LE BAS, N. C. MITCHELL, M. L. SOMERS, L. M. PARSON, and PH. PATRIAT / GLORIA Image Processing: The State of the Art   21

J. M. REYNOLDS / High-Resolution Seismic Reflection Surveying of Shallow Marine and Estuarine Environments   41

C. M. R. ROBERTS and M. C. SINHA / A Fixed Receiver for Recording Multichannel Wide-Angle Seismic Data on the Seabed   49

M. C. SINHA, P. D. PATEL, M. J. UNSWORTH, T. R. E. OWEN, and M. R. G. MacCORMACK / An Active Source Electromagnetic Sounding System for Marine Use   59

Q. HUGGETT / Long-Range Underwater Photography in the Deep Ocean   69

PART II: SAMPLING TECHNIQUES

P. P. E. WEAVER and P. J. SCHULTHEISS / Current Methods for Obtaining, Logging and Splitting Marine Sediment Cores   85

W. R. PARKER and G. C. SILLS / Observation of Corer Penetration and Sample Entry during Gravity Coring   101

M. A. STORMS / Ocean Drilling Program (ODP) Deep Sea Coring Techniques   109

R. B. KIDD, Q. J. HUGGETT, and A. T. S. RAMSAY / The Status of Geological Dredging Techniques   131

J. WHITE / The Use of Sediment Traps in High-Energy Environments   145

P. J. SCHULTHEISS / Pore Pressures in Marine Sediments: An Overview of Measurement Techniques and Some Geological and Engineering Applications   153

List of Contributors   169

Editorial Introduction

This collection of papers originates from a meeting organized in May 1988 at the Geological Society, London, under the auspices of its Marine Studies Group. The meeting was concerned with reviewing the present state-of-the-art of marine geological and geophysical sampling and surveying techniques.

The pace of scientific exploration of the ocean basins has increased dramatically over the past few decades in response to interest in the global tectonic processes which control their long-term evolution and the regional and local sedimentary and tectonic processes which shape them, as well as more practical questions such as the nature and extent of offshore mineral resources, problems of waste disposal at sea and the response of sea level to global climatic change. Marine geological research commonly requires that investigations be carried out in a difficult and often hostile environment. Observations must be made and instruments controlled at the sea floor, often many kilometres below the observer or recorder at the sea surface. In marine geology, more than in most other branches of earth science, major advances have depended on corresponding developments in available technology.

Although a number of general marine geological texts have been published in recent years, few of these address the specific problems of data acquisition across the broad spectrum of marine geological surveying. We felt that the pace of change in recent years has been so great that there was a clear need for a text which reviews recent developments in this field in sufficient detail to be of value to practicing marine geologists as well as to others requiring a more general appreciation of marine geological surveying techniques. This collection of papers attempts to meet this need.

The first requirement of any study of the ocean floor is to know the position of the survey vessel and the sampling/surveying instrument with the maximum possible accuracy. The first paper, by Robertson, provides a summary of the position-fixing techniques, for both ship and instrument, that

are in current use on board UK research vessels. Marine geological exploration requires information under three further headings: (i) the "shape" of the sea floor, (ii) the nature of the rocks and sediments which lie at its surface, and (iii) the nature of deeper structures. Studies of the shape of the sea floor (bathymetry) are based primarily on echo sounder and side-scan sonar surveying. Technology in this field has seen major advances over the past two decades, with the development of new ceramic materials to provide more efficient and powerful transducers, the increasing use of digital data processing techniques to improve the quality of the signal from the sea floor, and the introduction of new design concepts to provide higher resolution records.

Notable advances in this field include the development of commercially available narrow-beam echo sounders, such as the Atlas Parasound system described in the paper by Grant and Schreiber, which provides greatly enhanced resolution compared with conventional broad-beam systems. Imaging of larger areas of the sea floor is achieved through side-scan sonar surveying. The paper by Searle et al. discusses recent developments in the use of the UK Institute of Oceanographic Sciences long-range side-scan sonar system GLORIA, particularly in the fields of image processing, accurate determination of backscattering levels (important in the use of these data for determining bottom type) and in the digital combination of individual records into composite mosaics. The GLORIA system sweeps the sea floor to distances of about 0.5 km either side of the research vessel. The resulting record provides a more-or-less continuous acoustic plan of the sea floor in which individual relief features are recognized from their geometry, acoustic reflectivity and the acoustic shadows which they cast. However, precise bathymetric data over the insonified area are not determined.

An alternative approach is that of "swathe sounding" in which a number of sharply focused sound beams making different angles with the vertical are transmitted simultaneously from a single transducer

Marine Geophysical Researches **12**: vii–ix, 1990.

array. Each beam provides a separate precise depth determination at a different range from the vessel's track and these are digitally combined to produce a contoured bathymetric map of the survey area. The US Sea Beam system utilizes this principle and has been widely used, particularly by North American marine research institutes, over the past decade or so. A more recently developed system, based on the same general principle is the Atlas Hydrosweep system, described in the paper by Grant and Schreiber.

On a smaller scale, information on the nature of the sea floor can be obtained from direct observations using submersibles or underwater television and photography. The paper by Huggett provides a review of recent developments in deep ocean underwater photography, with a particular emphasis on the problems of light scattering and data retrieval.

The second major requirement of marine geological surveying is to obtain samples of the rocks and sediments which form the upper layers of the sea bed, in order to investigate geologically recent processes in the ocean basins. The four techniques used to obtain such samples are grab sampling, dredging, coring and ocean drilling. The most widely used technique for sampling soft sediments is coring. The paper by Weaver and Schultheiss reviews the different kinds of sediment corer in current use and their various advantages and disadvantages. It also discusses the techniques that are used for logging various physical properties of the sediments, splitting the cores, describing them and taking sub-samples. Penetration of the corer into the sea floor commonly compacts the sediment, so that the thicknesses of sedimentary units observed in the core are not necessarily representative of the true thicknesses of the geological units beneath the sea floor. Recent experiments, in which penetration of the corer have been monitored acoustically are described in the paper by Parker and Sills, which provides important new insights into the question of the fidelity of the geological record observed in sediment cores.

Harder rocks, particularly igneous and lithified sedimentary rocks normally cannot be sampled by simple gravity-powered corers and the technique commonly employed for sampling such rocks at the sea floor is that of dredging. Despite the widespread use of this technique, it is poorly documented in the literature. The contribution by Kidd *et al.* reviews the different types of dredge in current use for scien-

tific sampling and the particular limitations and applications of each type.

Perhaps the greatest advances in scientific sampling of ocean-floor rocks, particularly those at depths in excess of a few tens of metres beneath the sea bottom, have been made through the development and application of rotary drilling and sub-bottom coring techniques by the Deep Sea Drilling Project and its successor, the Ocean Drilling Program, over the past two decades. The review by Storms provides a detailed account of the particular problems involved in the efficient recovery of core samples by deep sea drilling and discusses recent advances in core recovery and low-disturbance drilling techniques.

In addition to obtaining samples of rocks and sediments from at and beneath the sea floor, marine geologists also require to obtain samples of the particles being deposited from the water column at the present time, in order to explore processes of sediment transport and deposition. This is achieved by the deployment of sediment traps. The paper by White reviews the different types of sediment trap that are in current use and the extent to which different designs provide a reliable measure of the true sediment particle flux to the sea floor.

Insight into geological processes occurring at or beneath the sea floor can be obtained from measurement of various physical and geotechnical properties of sea-floor sediments. One such property which has received attention in recent years is pore-pressure. This property can provide important information, for example, on hydrothermal circulation processes in the oceanic crust. Techniques for measuring this property and examples of the value of the data obtained are reviewed in the paper by Schultheiss.

The third major requirement of marine geological surveying is to obtain information on the structure of the sea floor to greater depths than can be explored by simple gravity coring and over wider areas than can be represented by isolated deep drilling. Such information is obtained by geophysical surveying, principally using seismic techniques. Multichannel seismic reflection surveying is widely used in the oil exploration industry and current techniques are well-described in the literature. The present publication focuses on two aspects of seismic techniques. The first is the application of high-resolution single-channel seismic reflection profiling to exploring the geology of shallow marine and estuarine environments, discussed in the contribution by Reynolds. The second is the particular problem of recording multichannel wide-angle seismic data on the continental shelf. The latter topic is addressed in the paper by Roberts and Sinha, which describes the principles and operational procedures used in a new system, the Pull-Up Multichannel Array (PUMA), together with results of preliminary deployments of the system on the UK continental shelf.

A relatively new technique of marine geological exploration is electromagnetic sounding. This provides information on the properties and distribution of conductive fluids within the oceanic crust. Such fluids include hydrothermal water and magma, so that electromagnetic sounding can provide important new insights into geological processes such as hydrothermal convection and magma emplacement at oceanic spreading centres. The paper by Sinha *et al.* describes a new approach to the problem of electromagnetic sounding at sea, utilizing an active source,

together with results of initial experiments which demonstrate the very considerable value of this new marine geological surveying technique.

This collection of papers cannot be claimed to be comprehensive in the sense of a textbook. However, we believe that the reader will gain a broad perspective of the state-of-the-art in many areas of marine geology through these papers and we express our thanks to our contributors for their efforts to provide this overview.

E. A. Hailwood
Oceanography Department,
University of Southampton

R. B. Kidd
Geology Department,
University College of Wales
Cardiff

September 1989.

PART I

Surveying Techniques

Deep Sea Navigation Techniques

K. G. ROBERTSON

Natural Environment Research Council, Research Vessel Services, No. 1 Dock, Barry, South Glamorgan CF6 6UZ, UK

(Received 27 April, 1989; accepted 1 September, 1989)

Key words: Deep sea navigation, position fixing, scientific surveying.

Abstract. Accurate navigation forms an essential part of all research at sea and the deep ocean imposes it's own unique problems. This chapter discusses several of the techniques in current use on the research vessels of the Natural Environment Research Council (NERC), concentrating on those systems which provide global navigation facilities, as opposed to the more localised, coastal aids. Whilst most of the systems rely on surface propagation of radio waves, the use of acoustics and sea-bed mapping instruments constitute accurate alternatives for some sub-sea applications.

Introduction

A wide variety of navigation aids is available to assist the scientist working in the marine environment. This chapter concentrates on those systems which can be employed in deep sea applications and, in particular, those currently used by the Natural Environment Research Council on the research vessels RRS Charles Darwin, RRS Discovery and RRS Challenger.

Although the main accent of this chapter is on marine geology, the techniques described are equally well suited to other branches of science at sea.

What is Navigation?

In the context of marine geology, navigation can perhaps be defined as the means by which one achieves the following objectives.

1. To locate oneself accurately at a point on the surface of the ocean, above the sea-bed feature under investigation, for example when locating a site at which to take core or grab samples, or to make underwater TV or photographic observations.

2. To manoeuvre a ship along a chosen track, whilst maintaining an accurate knowledge of one's position relative to known sea-bed features. For example, underway geophysical studies using towed air guns, hydrophone streamers, side scan sonar, magnetometers and shipborne gravity meters.

3. To manoeuvre accurately a submerged package along a chosen track on the sea-bed, for example, guiding deep towed instruments and Remote Operated Vehicles. This technique is also applicable to refining the objectives described in 1. above.

It will be obvious that success will often depend upon using a combination of navigation techniques, a knowledge of any previous work carried out in the same region and access to the relevant, detailed charts and geological maps, where they exist.

Navigation Techniques

Techniques used by NERC have been refined over many years as technological advances have given us new tools with which to operate. With the aid of modern computing systems, installed on each of the NERC ships, we are able to combine the outputs of many devices and apply quality control parameters that can reduce still further the individual instrument errors and produce high quality, annotated track charts in near real time.

The suite of navigation instruments one would normally expect to be present on board when working in deep water would include the following.

> TRANSIT Satellite Receiver.
> GPS Satellite Receiver.
> Gyro Compass.
> Electromagnetic and/or Doppler speed sensors.

For more specialised applications other instruments may also be included, such as:

Radar Transponders.
Microwave Ranging Systems.
Acoustic Transponders.

Looking ahead, these may be supplemented in the future by such instruments as:

Swathe or Multi-Beam sounders.

Each of the techniques listed above is described in more detail in the following sections.

TRANSIT

The TRANSIT, satellite based system has been the mainstay of deep sea navigation within NERC for almost two decades. A number of polar orbiting satellites circle the earth every 107 minutes, at an altitude of 600 nautical miles.

The orbits trace a pattern like the sections of an orange, wide at the equator and converging at the poles (Fig. 1). However, the orbits do not rotate with the earth so they suggest a sort of cage, inside which the earth turns on it's axis. Consequently, each point on the earth's surface passes under each of the orbits approximately twice a day.

Each satellite transmits information, as a function of time, about it's position relative to the centre of the earth. By measuring the change of Doppler frequency of the received signals as the satellite approaches, passes and recedes, it is possible to fix the position of the ship relative to the satellite and hence the precise position on the surface of the earth. However, as there is a relatively small number of satellites and they can exhibit a degree of "bunching", it is possible to encounter delays of up to a few hours between position fixes.

In order to update one's position regularly it is necessary to make a series of close estimates, based on the ship's heading from the gyro compass and speed from the electromagnetic or doppler log, until another satellite transmission is received. This procedure is described as 'Dead Reckoning' and is open to error from any inaccuracies in the calibration of the sensing devices. Cumulative errors can result in apparent jumps in one's assumed track when another reliable spot fix is obtained (Fig. 2). With the use of modern computing techniques the data can be reprocessed to minimise the uncertainties.

Although now beginning to be superceded by GPS, the TRANSIT system remains a useful tool and will continue to be operational for many years yet.

If one has the opportunity to accumulate a series of spot fixes on a stationary target it is possible that an accuracy of 100 metres, or better, can be expected. For a moving ship this will be downgraded since the reliability of a fix will depend upon the quality of the data coming from the satellite, the characteristics of its orbit relative to the ship, the forward velocity of the ship and any ship motion induced accelerations.

Typical accuracy in the deep sea is in the region of 100 m to 1 nautical mile, dependent upon the parameters described above. After prolonged periods of 'Dead Reckoning' the accumulated errors can create an uncertainty of 2 km or so but this can vary greatly.

GPS (NAVSTAR)

Undoubtedly, the navigation system upon which NERC will become increasingly reliant is the Global

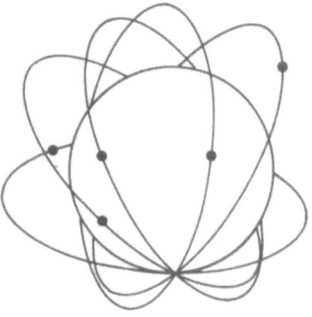

Fig. 1. Transit orbits.

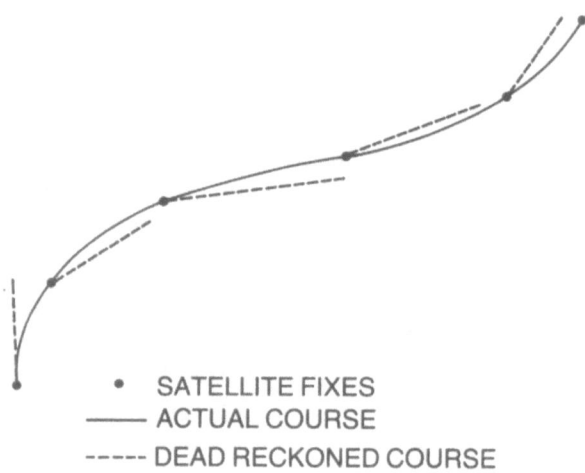

Fig. 2. The effect of 'Dead Reckoning' procedures.

Positioning System, or GPS. This consists of a constellation of satellites, normally referred to as Space Vehicles, or SV's, orbiting the earth every 12 hours, associated ground stations and user equipment.

When completed, there will be 18 SV's in 6 different orbital planes, inclined approximately 63° to the equator and at an altitude of 20,000 km (Fig. 3). The system will provide for three dimensional position, horizontal and vertical velocity and extremely accurate time, on a continuous, worldwide basis. All SV's were due to be launched by the end of the 1980's but the delay in the Space Shuttle programme has put back full deployment by several years.

The basic measurement made by the user's receiver is that of the apparent propagation time of a timing mark from the SV to the receiver antenna. By making this measurement from three SV's the user's Latitude and Longitude can be determined. Furthermore, by measuring the rate of change of range, the user's speed and heading can be computed.

The satellite positions in space are measured to within a few metres and the satellite clocks are calibrated to a few nanoseconds so that computation of position on the surface of the earth is possible to within a few metres.

However, each satellite transmits two codes, the P, or Precision code and the CA, or Coarse Acquisition code. To gain access to the full precision of the system a user would need a knowledge of the P code. Commercial users have access only to the CA code and as a consequence the likely order of precision is in the region of 100 metres.

The advantages are that computation of position will be continuous and without the cumulative errors associated with Dead Reckoning procedures. With the added facility of determining velocity and heading it is obvious that GPS has a great potential as an aid to marine science.

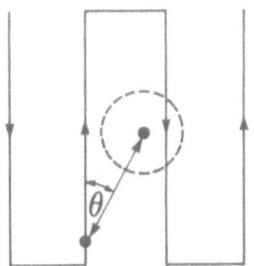

Fig. 4. Radar transponder. Typical grid pattern survey.

RADAR TRANSPONDERS

Although not used by NERC on a regular basis any more this technique was employed for many years and much useful work was accomplished with it.

An active radar transponder is attached to the top of a purpose made buoy which contains a sealed compartment to house the battery power supplies. The buoy is tethered to an anchor weight by a length of light steel wire and its surface position is determined as accurately as possible by means of time series of satellite fixes. Surveys are then conducted whilst taking frequent readings of the range and bearing of the transponder, using the ship's radar (Fig. 4). A series of buoys could give coverage of a larger area or the chance of greater precision, if needed.

Unfortunately there are several practical limitations which restrict the usefulness of the technique, especially in deep water. The buoy remains subject to movement from surface currents and winds, thus creating a degree of uncertainty with regard to its absolute position at any time (Fig. 5). In addition, the time taken to obtain enough satellite fixes to determine the buoy's position with sufficient accuracy can become prohibitive.

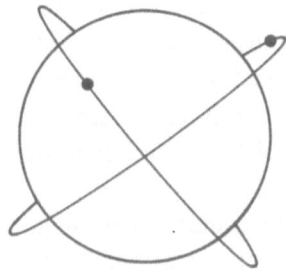

Fig. 3. GPS orbits (only 2 are shown for clarity).

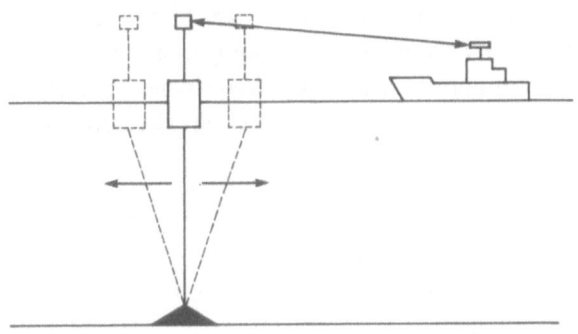

Fig. 5. Radar transponders. The effect of tidal and surface current movement.

The practical limit to the range from a single buoy is about 20 miles and is dependent on sea state conditions but nevertheless the technique allows almost continuous positioning and is still used for some shallow water applications.

MICROWAVE RANGING

With a normal operating range of some tens of kilometers this technique would not seem applicable to deep sea work, at first sight. However, geophysicists will be aware that there are certain tasks which demand the use of two ships working together. Commonly one would be operating the acoustic energy sources while the other would be listening and recording with a multi-channel hydrophone system.

When the two ships are traversing the same track, in the same direction, the Microwave Ranging system provides a constantly updated and accurate measure of their separation, thus allowing one to increase, or decrease its forward speed to maintain the distance between them to within relatively fine limits (Fig. 6).

When the ships are traversing the same track but in opposite directions, it is normal to maintain a steady speed and use Microwave Ranging to measure accurately their increasing, or decreasing separation (Fig. 7).

By this means NERC has frequently achieved a precision of a few tens of metres over separation ranges of several tens of kilometres.

ACOUSTIC TRANSPONDERS

All systems described so far can be grouped into the category of radio navigation aids. That is, they operate by transmitting to a receiver through the air, at high frequencies.

If, however, it is necessary to navigate a vehicle or towed device beneath the surface of the ocean one must look to a different technology, based on acoustic communication at lower frequencies. The equipment used for this type of application would

Fig. 6. Microwave ranging used to maintain range of pair of seismic survey vessels.

Fig. 7. Microwave ranging used to maintain separation of vessels.

commonly consist of a number of underwater acoustic transponders, a surface transducer to interrogate them and instruments to translate the data received into positional information.

Having navigated the ship to the area of interest by the use of conventional surface methods the first step is to lay a pattern of transponders on, or close to, the sea bed. When interrogated from the surface the transponders in range will reply and the time taken for this reply to be received will be a function of the slant distance from the ship. As replies are received from a number of transponders in the pattern it is possible to compute the ship's position relative to the sea bed instrument array.

This technique can be extended to cater for one or more towed, or free swimming vehicles. For example, if interrogation of sea bed transponders is conducted through a transponder mounted on the submerged vehicle then its position, relative to the array, can be determined. A combination of surface and submerged transducers allows one to track both the towed vehicle and the ship (Fig. 8).

With the use of conventional surface navigation aids the position of the sea bed transponder array can be translated into geographic coordinates. All such systems include calibration routines to achieve this end if desired. Under some circumstances it is possible to achieve relative positional accuracy of a metre or better but as with all navigation systems, there are sources of possible error.

It is essential to establish a precise knowledge of local variations in the velocity of propagation of sound in water. Calculation of range from the measurement of transit time can be heavily dependent on this parameter if inconsistencies are encountered. Internal refraction in the water column can alter the range over which reliable acoustic communication can be expected.

Line of sight conditions must normally be established for correct operation. In the presence of sea bed projections it may be necessary to moor the transponders at distances above the ocean floor

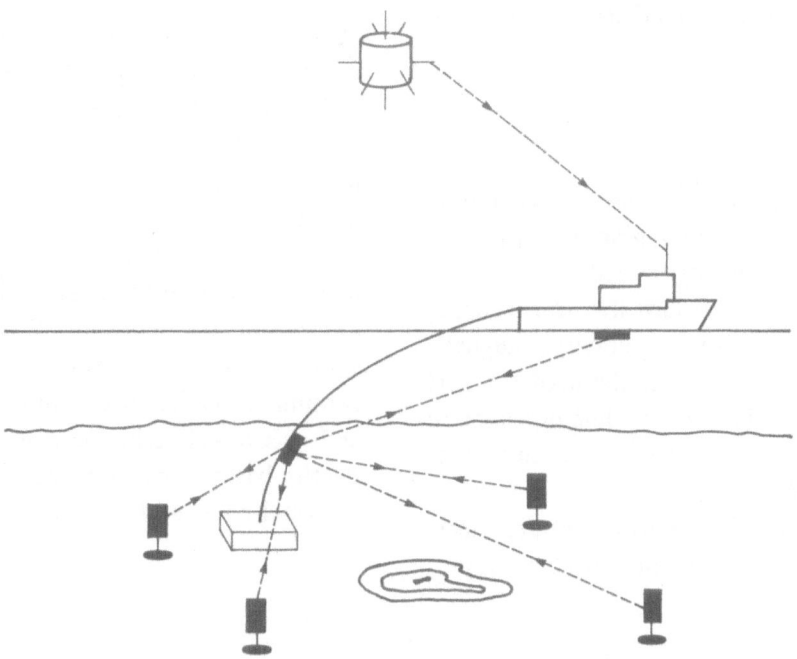

Fig. 8. Acoustic transponders. A combination of acoustic transponders and satellite navigation allows tracking of the surface ship and sub surface vehicle.

and hence they become more likely to be influenced by currents. Thus distances relative to each other and to surface coordinates have a degree of uncertainty.

The accuracy to which one can translate positions relative to a transponder pattern into positions relative to geographic coordinates depends on the accuracy of the radio navigation aids with which the ship is equipped and how well they are operating at the time of system calibration.

Acoustic transponders are not limited to use in the manner described above. The techniques of Long Baseline, Short Baseline and Ultra Short Baseline surveying are well documented elsewhere. Applications are numerous but two areas which could be of interest to the marine geologist might be:-

The precise positioning of a corer within a sea bed array (Fig. 9).
The tracking of arrays of seismic energy sources (Fig. 10).

SWATHE SOUNDERS

Swathe, or Multi-Beam sounders are discussed elsewhere in this volume. Their capability to map, in great detail, large areas of sea bed, coupled with processing possibilities of constructing contour plots

and representations of the three-dimensional topographic characteristics offer the scientist an opportunity for positioning his sampling device with great precision.

The production of such detailed pictures will facilitate very accurate selection of sites or survey tracks

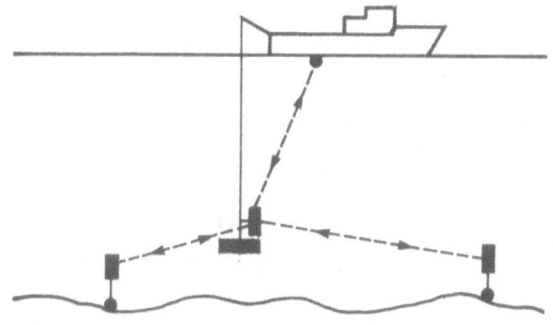

Fig. 9. The use of acoustic transponders to position a corer.

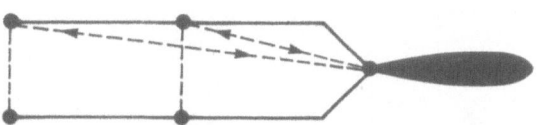

Fig. 10. The use of acoustic transponders to monitor the towing characteristics of near surface air gun or hydrophone arrays.

and if near real-time verification is obtainable this will establish a high level of confidence that one is exactly where one wished to be.

The Role of Computers

As mentioned earlier, intelligent use of computing can improve navigation by combining the outputs of several devices and eradicating, or reducing, sources of errors. The ability to produce live track charts, annotated with depth, position, gravity, magnetic field, station number etc., reduces the need for post cruise processing and allows for on-line decisions to be made regarding the conduct of the survey or sampling programme.

By relating all measurements to a time reference, data can be correlated and assessed more readily.

Conclusions

Many aids to navigation are available to the scientist at sea. This paper has concentrated on those techniques which are most applicable to deep water work, where one is usually distant from the localised hyperbolic systems and alternative Radar targets.

There is little doubt that the continued development of the GPS system will have the greatest impact on the work of NERC for all types of surveys. The combination of this high quality and continuous global positioning with the more local and specialised techniques of underwater acoustics, swathe bathymetry and microwave ranging offers a powerful package to the marine scientist.

Modern Swathe Sounding and Sub-Bottom Profiling Technology for Research Applications: The Atlas Hydrosweep and Parasound Systems

J. A. GRANT and R. SCHREIBER

Krupp Atlas Elektronik Gmbh, Sebaldsbrücker Heerstraße 235, D2800, Bremen 44, FRG

(Received 27 April, 1989; accepted 1 September, 1989)

Key words: Swathe sounding, narrow-beam sub-bottom profiling, acoustic surveying.

Abstract. This chapter describes two separate but complementary research echosounder systems originally developed by Krupp Atlas Elektronik GmbH for the new German oceanographic research vessel *Meteor*.

The *Hydrosweep* is a Hydrographic wide-swathe sweep survey echosounder for both shallow and deep water applications providing accurate bathymetric surveys and terrain-following navigation capabilities.

The *Parasound* system is a hull-mounted dual channel parametric narrow-beam deep sea survey and sub-bottom profiling echosounder enabling particularly high vertical and horizontal resolution of seabed features.

Swathe Sounding Systems

The experimental work of Colladon and Sturm (1827) to determine the speed of sound in water, laid the foundations for the eventual development of the echosounder, which, a century later, had replaced the leadline as the traditional method of sampling water depth. The next 50 years saw advances in both acoustic and electronic engineering which led to the modern precision echosounders and the ability to acquire and evaluate depth data to a high order of accuracy.

The conventional precision echosounder, however, can only measure the water depth immediately below the survey vessel, and in order to achieve satisfactory coverage of the seabed it is usually necessary to survey sets of parallel sounding lines (Grant, 1985). The number and spacing of these lines will depend upon a number of factors including the scale of the survey and its purpose. Even with closely spaced survey lines it is still not possible to guarantee full coverage of the seabed, and isolated features that lie between the lines may be missed. Various approaches have been used in trying to solve this problem including, with some success, echosounder systems such as the Atlas BOMA system using multiple transducers mounted on booms, and the deployment of survey launches sailing on tracks parallel to that of a mother ship (Stenborg, 1987).

The 1970s saw the development of the first commercial swathe bathymetry systems (de Moustier, 1988; Wentzell and Ziese, 1988). These were either of the multibeam echosounder type, such as Sea Beam (Farr, 1980) or the less accurate, towed interferometric sidescan type with swathe bathymetry capabilities, typified in the USA by SeaMARC (Blackinton *et al.*, 1986), and in the UK by Bathyscan (Cloet and Edwards, 1986).

It has been shown that Sea Beam, with its comparatively narrow swathe width and restricted capabilities in shallow water, suffers from a number of errors resulting in incorrect depth determination, which may cause artifacts to be observed in the data collected (de Moustier and Kleinrock, 1986). Modern oceanographic research increasingly requires depth data to be collected over large areas of the seabed with great precision and efficiency.

The multibeam system *Hydrosweep* is designed to meet this need, providing the geologist and hydrographic surveyor with the capability of carrying out accurate, wide-swathe surveys in water that may be as shallow as 10 m or as deep as 11 000 m.

At the end of December 1987, the Atlas *Hydrosweep* on board R.V. Meteor was used to survey

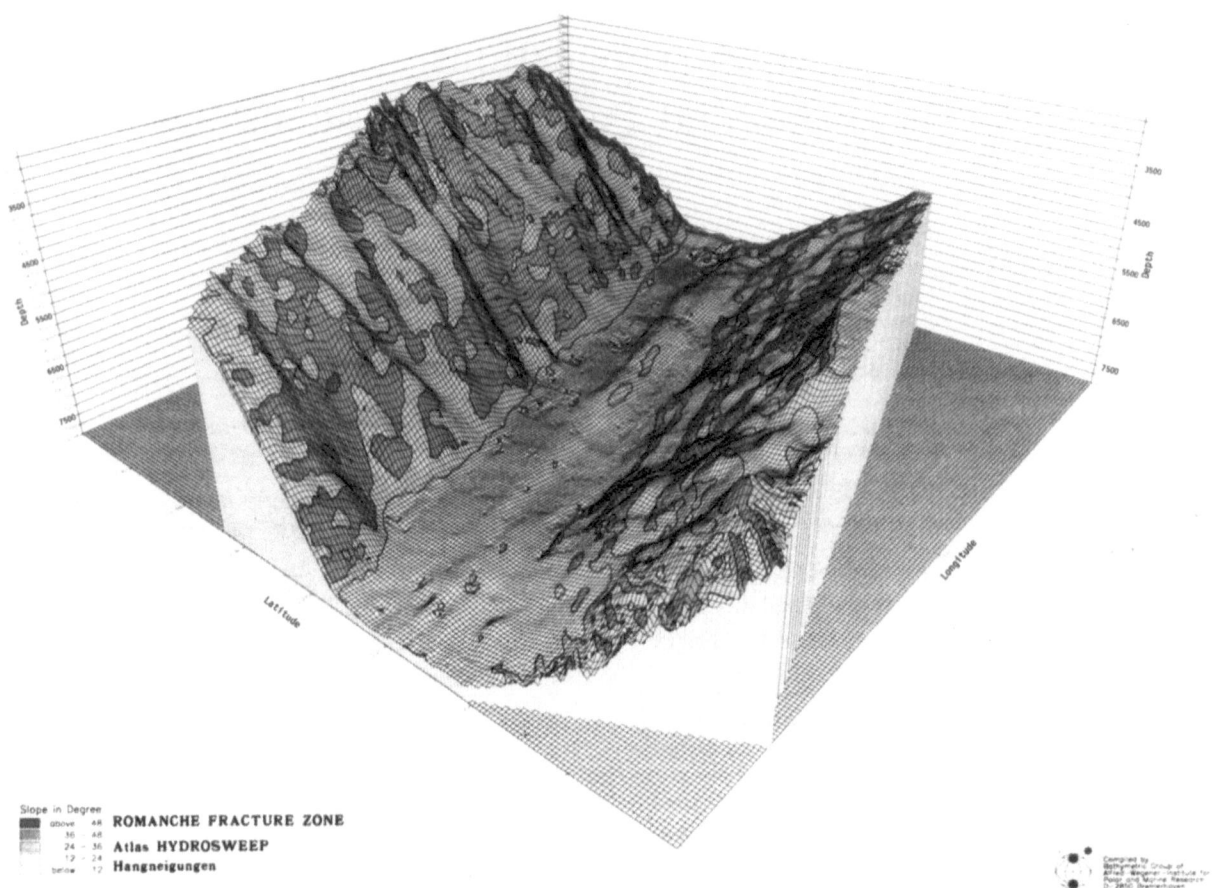

Slope in Degree

above   48   **ROMANCHE FRACTURE ZONE**
36   48
24 - 36   **Atlas HYDROSWEEP**
12 - 24
below   12   **Hangneigungen**

Fig. 1. Part of the Romanche Fracture Zone, shown as a perspective representation of a three dimensional terrain model, generated during post-processing at the Alfred Wegener Institute, Bremerhaven.

the Romanche Fracture Zone (Siedler *et al.*, 1987). The central area of this trench zone is about 10 km wide with water depths of up to 7900 m, while the slopes at the sides of the trench have inclinations which can be greater than 45% (Fig. 1). *Hydrosweep* reached the bottom of the trench with its entire fan of sound beams, a swathe width of more than 14 km (Schenke, 1988).

The Atlas Hydrosweep System

The *Hydrosweep* system is characterized by the very wide area which is covered by the 59 pre-formed beams (PFBs) transmitted in a fan from the hull-mounted transducers. The outer-beams may be radiated at angles of 45° relative to the surface, thus providing the capability of achieving a total swathe width equal to twice the water depth.

As with conventional vertical echosounders, variations of sound velocity in the water column influence the travel time of the acoustic pulse and therefore the

system accuracy. In multibeam systems with large swathe angles there is an additional effect on the divergence of the slant beams which may result in unacceptable errors in measurement, especially by refraction of the outermost beams. For this reason other swathe systems have had to make use of sound velocity probes to overcome the problem. This is time-consuming and therefore not cost-effective. *Hydrosweep* makes use of a unique, patented calibration process which takes place while the vessel is underway (Fig. 2).

During the survey all vertical depth values are measured by the centre beam and stored. These vertical depth determinations are, of course, not subject to slant error. At regular intervals the fan of beams is rotated through 90° to transmit along the ship's longitudinal axis and the resulting profile is measured and stored.

In an iterative process the two profiles are compared and a correction factor is applied to produce a

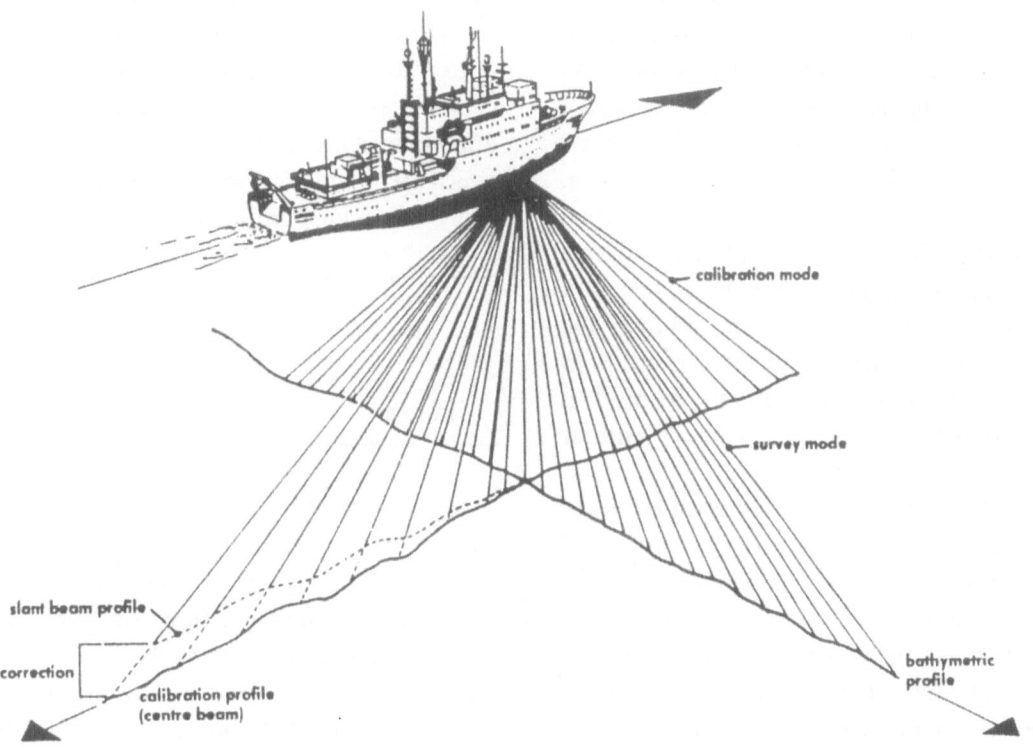

Fig. 2. Atlas Hydrosweep calibration procedure.

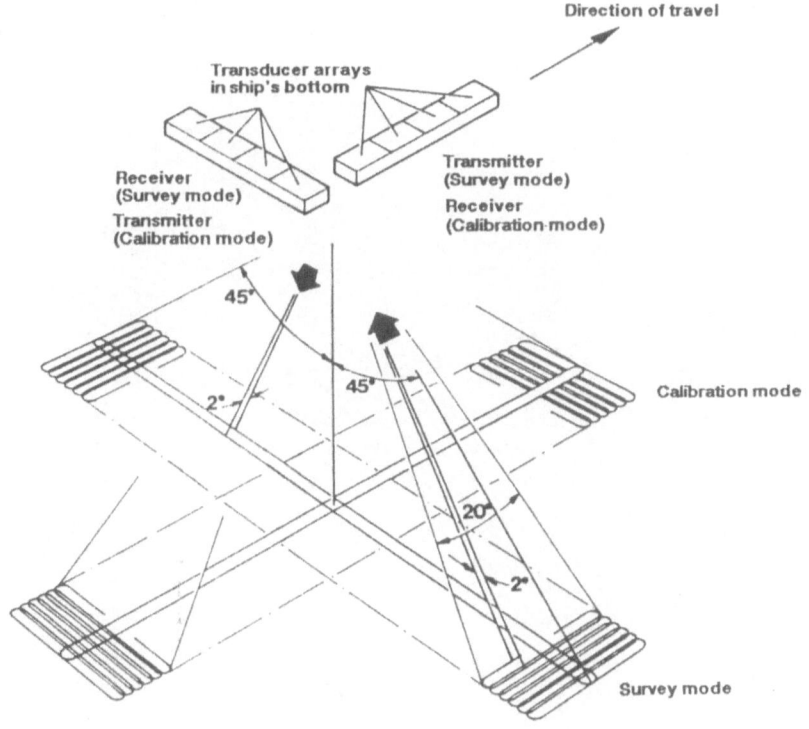

Fig. 3. Atlas Hydrosweep Transducer beam patterns.

'best fit' result. This calibration factor includes, among other parameters, the average sound velocity in the water column, and may be applied as a correction to both the slant and vertical beam travel time in the survey mode. This process is repeated automatically at regular intervals as the vessel travels along its track (Schreiber and Schenke, 1989).

Analysis of data from system trials have shown that the residual depth error may be reduced to better than 0.4% of water depth. In addition, the *Hydrosweep* calibration process makes it possible to sail from shallow to deep water without the need to deploy sound-velocity probes, and therefore without losing valuable operation time.

Hydrosweep uses two identical, solid-state piezo-electric transducer arrays, each array consisting of replaceable modules (Fig. 3). One of the arrays is installed along the longitudinal fore and aft axis of the vessel and during the survey mode is used as the transmitter, while the second array is set at right angles to the first and acts as the receiver. When the

system switches over to the calibration mode, the functions of the transducers are exchanged, and the fan beam rotates through 90°.

In order to maintain shallow and deep water capabilities the appropriate transmission mode is automatically selected by the system. In depths between 10 m and 1000 m, short pulse lengths and low energy emissions are used.

depth m	transmission mode	beam width
10–100	omnidirectional	90° × 4.5°
100–1000	omnidirectional	90° × 1.9°

In water depths greater than 1000 m, longer pulse lengths and higher energy are emitted using a Rotational Directional Transmission mode (RDT). The transmission is made in three steps, one centred on the vertical and the other at the 36° points with time separation between the pulses being kept to a minimum to ensure that the optimum sounding rate is

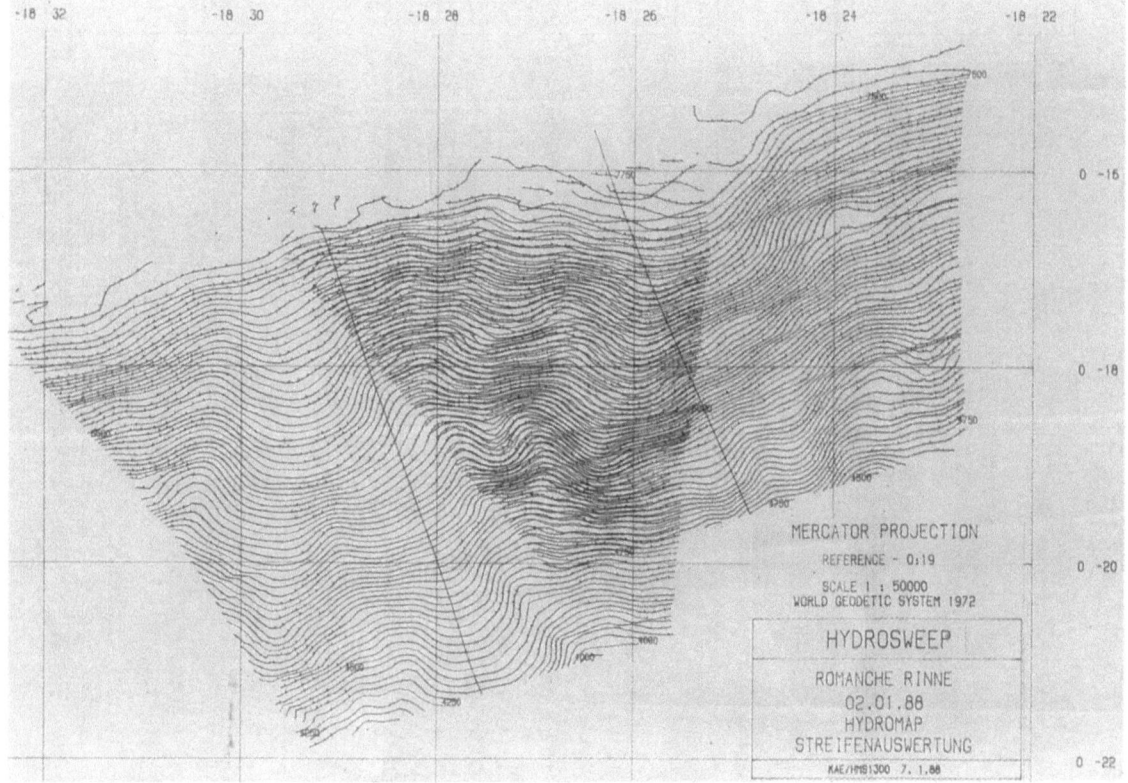

Fig. 4. Contour map of the Romanche Fracture Zone derived from Hydrosweep surveys. Overlapping parts of two separate surveys show excellent correspondence.

maintained. Using RDT in this way ensures that the system achieves uniform insonification across the swathe. The pulse length will vary according to depth, and is separately controlled on the vertical beam and on the outer sectors.

The so-called 'omega-effect' was experienced with earlier multi-beam systems when descending a steep slope, (de Moustier and Kleinrock, 1986). In such a situation the intensity of an echo generated by side lobes from further back up the slope may be comparable to the vertically reflected return from directly beneath the ship. The tracking gate within the system may then fail to follow the actual bottom, but will lock on to the false shallow depth produced by the side lobe echo and lag behind the changing slope. The false echo can produce a contour which in classic cases resembles the Greek letter omega (Ω).

The transmission pattern of *Hydrosweep* has been modified to neutralize this effect and during the *Meteor* survey of the Romanche Fracture Zone,

specific test measurements were made in the transition area between the deep sea floor and the steep sides of the trench (Fig. 4.) and no 'omega-effect' was observed (Tyce, 1987; Schenke, 1988).

The system is operated at a frequency of 15.5 kHz, which was chosen to achieve two objectives. On the one hand, this frequency is low enough to achieve the desired depth range, giving *Hydrosweep* with its highly efficient narrow beam transducers the capability to reach the full ocean depths in excess of 10 000 m, while the low noise signal processing techniques provide a precise determination of the depths from the 59 return signals.

On the other hand, the frequency is high enough to secure good vertical resolution, even in shallow water, and the signal is reflected from the true bottom, rather than penetrating the sediment layers to the sub-bottom surfaces. The chosen frequency is also high enough to achieve a good "Raleigh" figure, which determines the frequency at which a bottom of

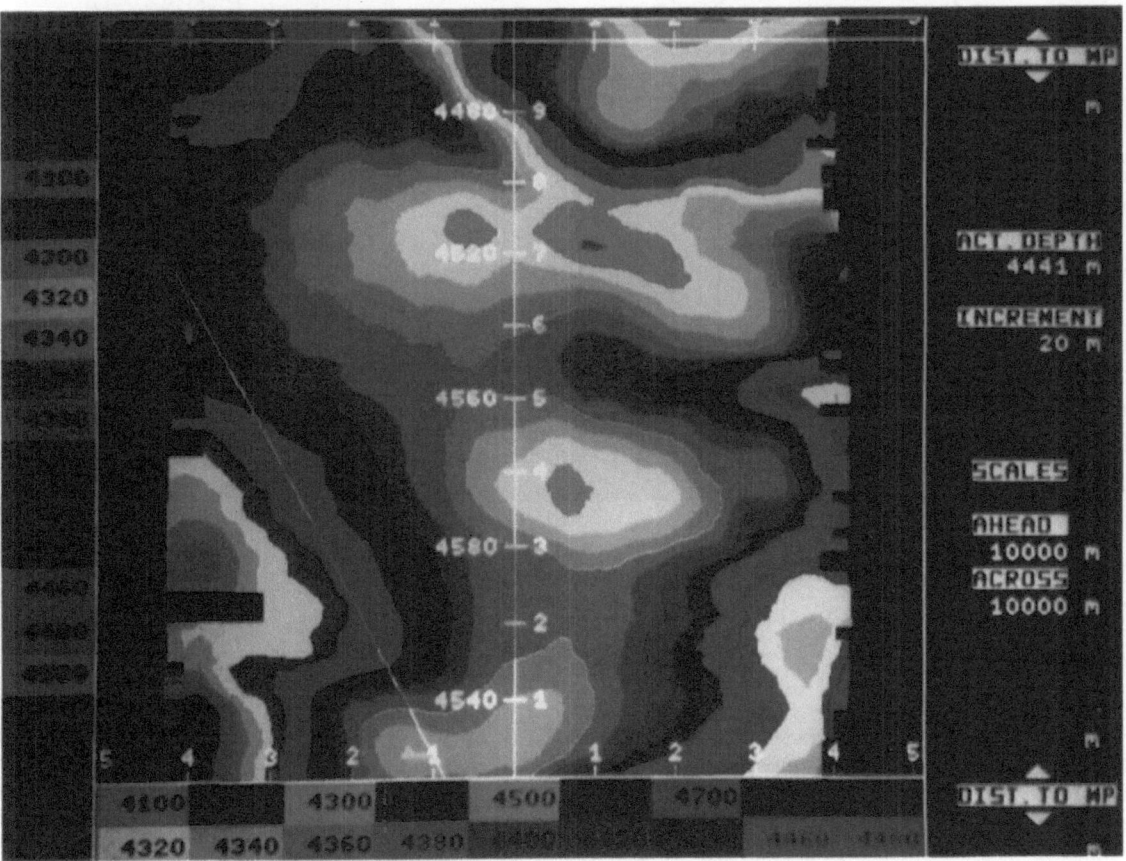

Fig. 5. Atlas Hydrosweep real-time colour graduated contour map. Vessel position is shown by the cross lines at the top of the display. The ships planned track is superimposed on the display and all other parameters such as depth, range across the display as well as the incrementation of the contours is shown.

a certain roughness is seen as a diffuse reflector (Urick, 1983). This will provide a better signal return at the outermost slant beams.

The return signal may be affected by a number of disturbing influences which may make discrimination of the true bottom echo difficult, and side lobe echoes may fall into the receiver time gates, falsifying the energy centre created by the main echo. Within the *Hydrosweep* System microprocessors exclusively allocated to echo discrimination carry out a routine which is repeated several times. An 'adaptive window' technique accurately determines the precise centre of

energy, which, by definition is the true bottom (Schreiber and Schenke, 1989).

It should be noted that the uncertainty in the horizontal position of a survey vessel currently exceeds the uncertainty in depth determination. In order to fully utilize the achievable levels of operational accuracy of a survey system a pre-requisite is that high-accuracy position fixing methods are employed together with compensation for heave, roll and pitch motion on the vessel.

The *Hydrosweep* system is controlled, and data collection monitored, from an operator's console.

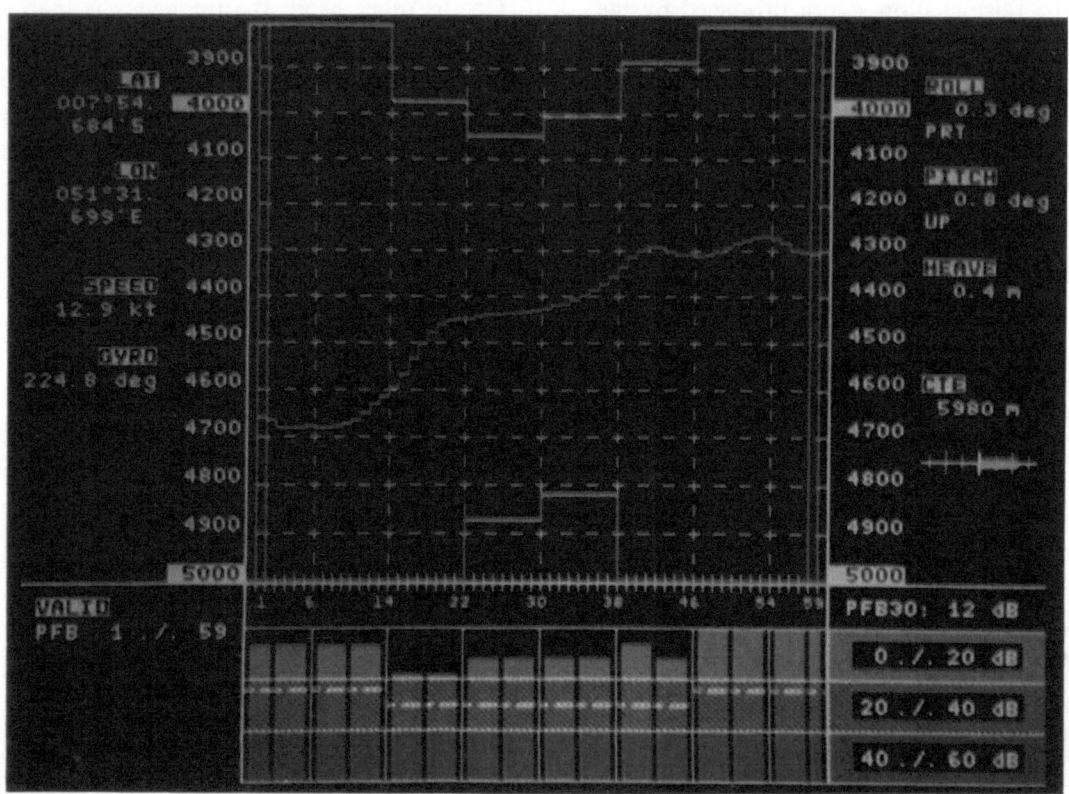

Fig. 6. Atlas Hydrosweep: profile and performance monitor display. The display is "steady", and its contents are overwritten anew with each receiving cycle.

Left
Navigation parameters.

Middle
Upper part cross profile display (during calibration: longitudinal profile):
(1) For each beam, the measured depth of water is displayed with a vertical linear scale.
(2) The "search window" for the depth digitizer is re-positioned automatically.
(3) The lateral boundary of the beams of which the data are to be recorded can be selected manually.
(4) The number of the beams are marked on this line.

Lower part for each of eight amplifier channels of the side beams, a bar graph is shown for the "attenuation", i.e. the relative reflectivity of the bottom.

Right
Annotation of dynamic parameters of the ship at the instant of transmission.

Dedicated function keys are used to call up or enter the required operation parameters on a monochrome alphanumeric display. Rapid system familiarization is made possible by the use of menu dialogue techniques, facilitating ease of operation even when changing from one function to the next.

Data are displayed in real-time on a second high-resolution VDU, either in the form of a graduated colour contour map (Fig. 5) or as a cross-section profile and performance monitor display (Fig. 6). Navigation during survey operations may be enhanced by being able to follow seabed contours, any deviation from pre-planned tracks being clearly shown. All data are recorded on digital tape, and may also be reproduced on isoline charts for immediate on-board evaluation.

Other sophisticated routines allow system testing in support of fault location, quality-control checks and total replay of survey data. This playback and simulation function also allows operator training to be carried out on-board the vessel, even in port (Tyce, 1987).

The Atlas Parasound System

The importance of being able to investigate the sedimentary structure of the seabed, whether for pure geological research or commercial offshore engineering purposes, has led to the development of various types of low-frequency echosounders which have been used to good effect as sub-bottom profilers.

These systems typically have used frequencies of around 3.5 kHz, the actual frequency depending upon the depth and nature of the bottom. While they can penetrate quite deeply, they do nevertheless suffer from a number of inherent disadvantages. The use of such low frequencies implies the employment of large transducers and wide beam-widths. This in turn leads to the generation of side echoes which interfere with the desired sediment echo, often rendering it virtually useless.

To overcome this problem, transducers have been mounted in deep-towed vehicles. However, this requires the use of special cables, winches, hardware and, of course, vessel operation at low speeds. The result is degraded data at high mobilization and operation costs.

The Atlas *Parasound* System combines a narrow beam, deep sea survey echosounder with a sub-

bottom profiler and represents a major new development in advanced hydroacoustic engineering. The techniques employed make it possible to achieve very high vertical and horizontal resolution, facilitating the detection of fine layers of sediment, while giving excellent bottom penetration from a hull-mounted transducer (Schreiber *et al.*, 1988).

The parametric principle has been established for some years and makes use of the non-linear acoustic properties of the water column (Berktay, 1965). When two signals at adjacent frequencies are radiated simultaneously into the water column at very high energy levels, they generate a secondary frequency which is equal to the difference between the two primary frequencies. The secondary frequency is only generated in the main part of the beam, where the highest energy levels occur. By this means a low frequency transmission, virtually free of side lobes, with a very narrow beam is produced (Fig. 7).

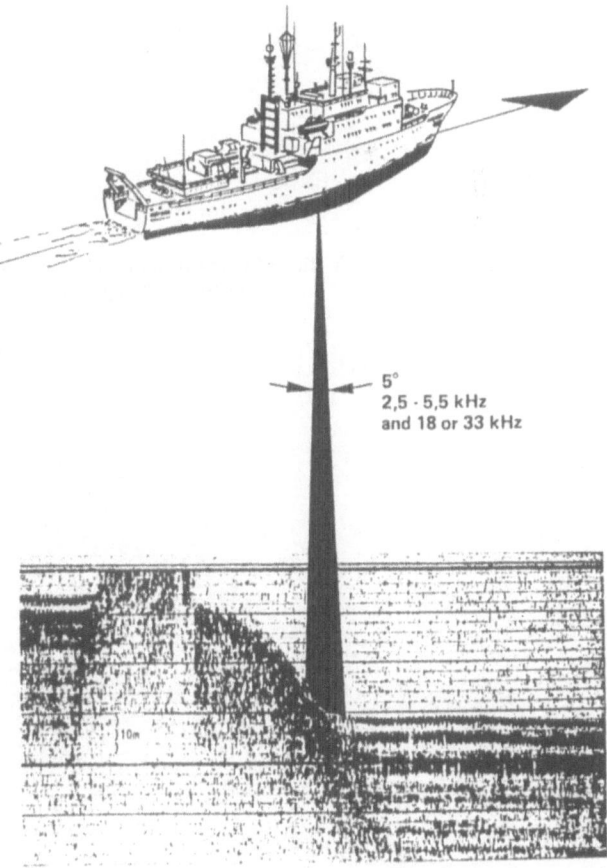

Fig. 7. Atlas Parasound—narrow beam echosounder and sub-bottom profiler.

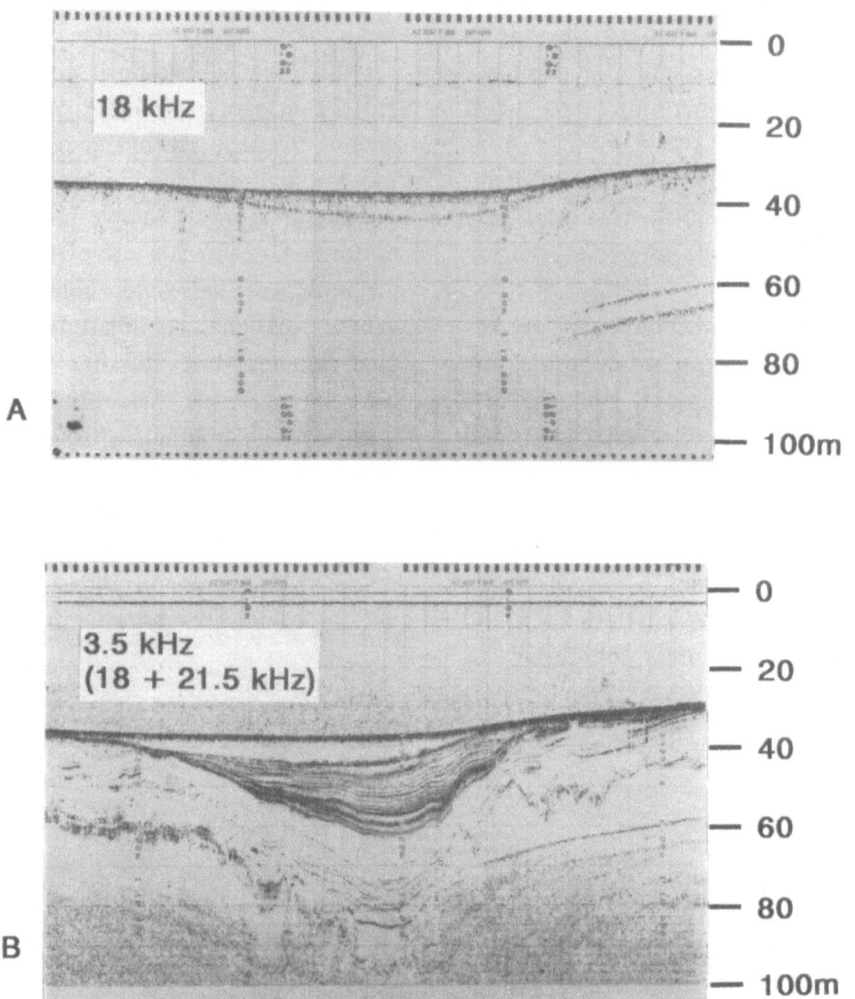

Fig. 8. The acoustic parametric effect. Comparison of seabed records made by R. V. 'Meteor' in the Kattegat. Echogram A at 18 kHz and B 3.5 kHz (18 kHz and 21.5 kHz transmitted simultaneously).

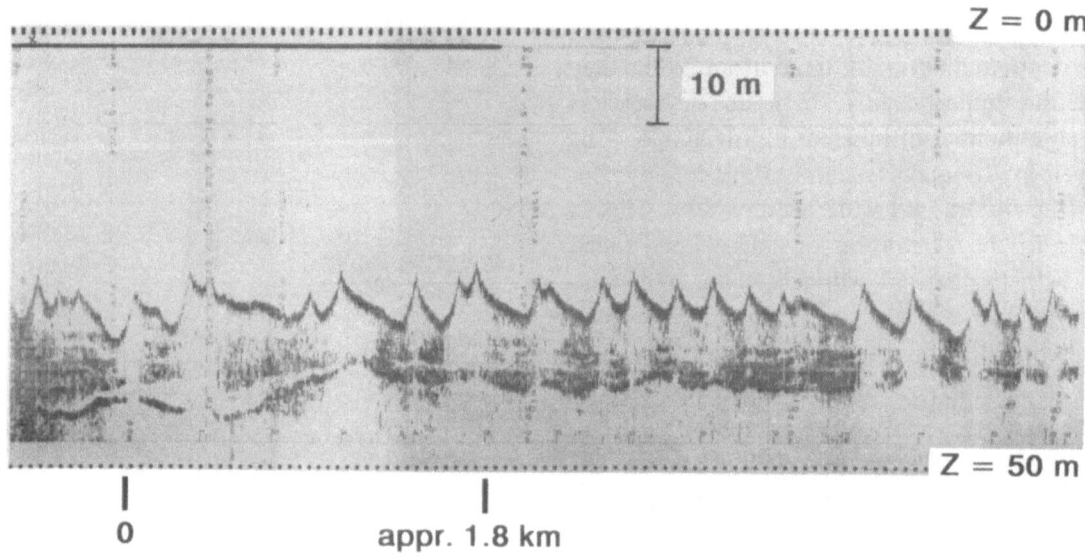

Fig. 9. Atlas Parasound record of giant active sandwaves with an elevation of up to 10 m at a water depth of approximately 30 m. (English Channel).

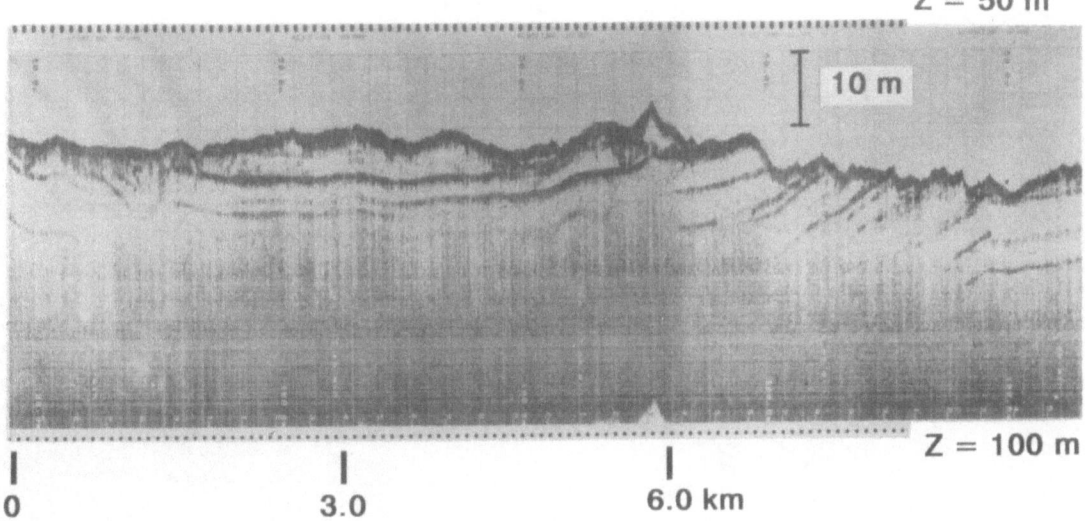

Fig. 10. Atlas Parasound record of alternating layers of slightly distorted hard and soft Mesozoic layers, with sand covering at water depth of 60–70 m (English Channel).

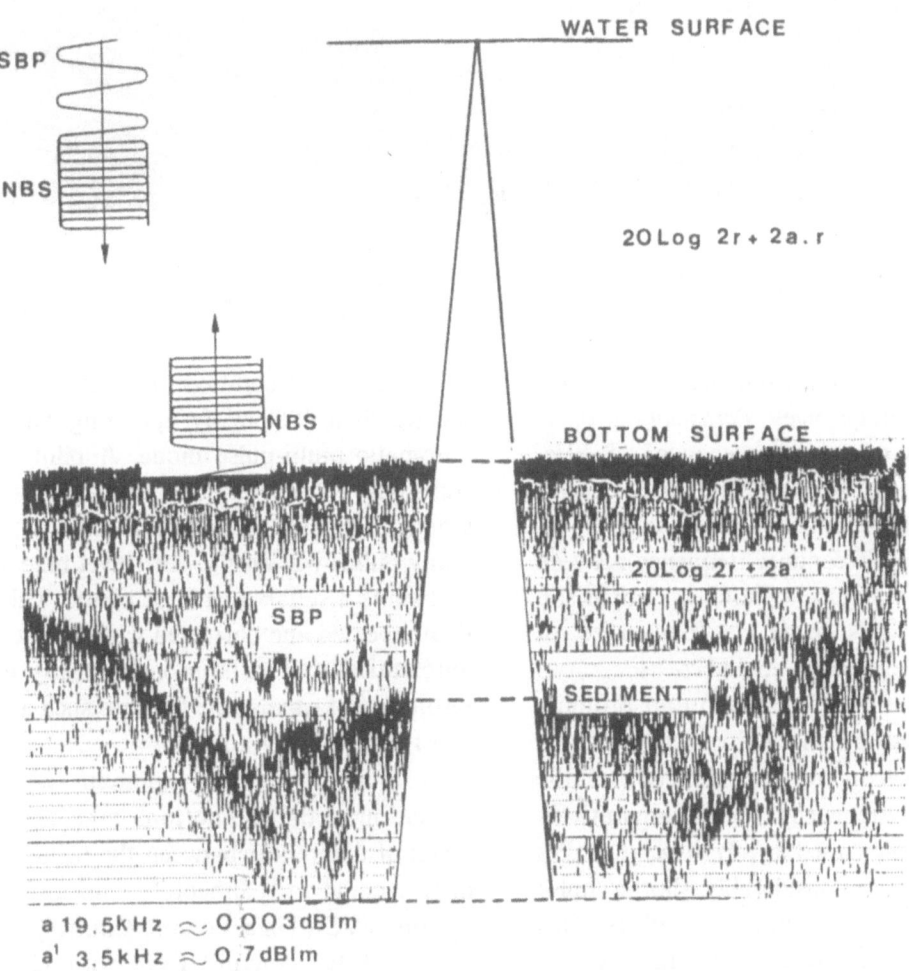

Fig. 11. Atlas Parasound—TVC behaviour in water and sediment. NBS – Narrow beam signal at the primary frequency acts as a pilot pulse and detects the true bottom. SBP – Sub-bottom profiler signal at the secondary frequency. r – Range in meters. a – Decibels km⁻¹.

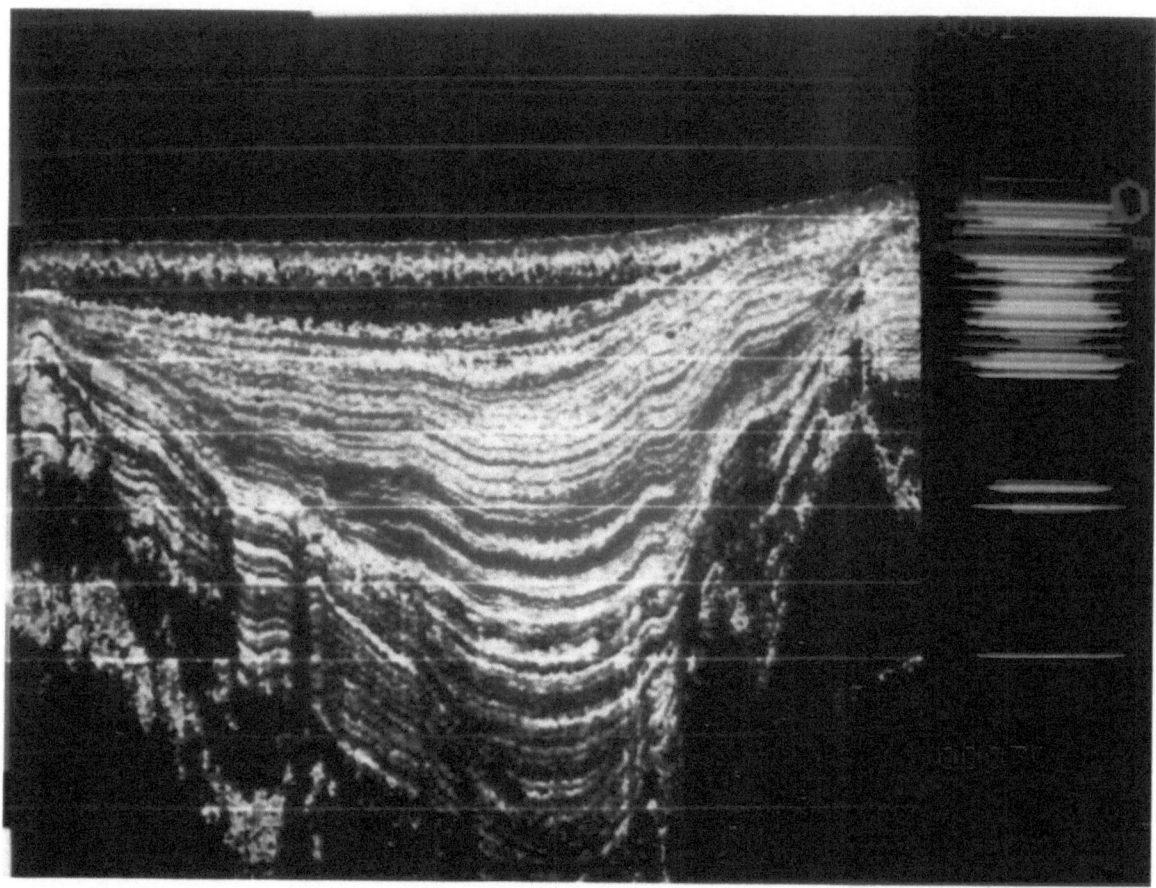

Fig. 12. Atlas Parasound—high resolution colour display. The main display on the left shows the reflected sediment layers at an appropriate scale (50 m in this case). The right-hand display is an expansion of the instantaneous signal.

In principle, any of the ultrasonic frequencies may be used as primary frequencies. *Parasound* is operated with one fixed primary frequency of 18 kHz, which allows the system to reach full ocean depths, with a second primary frequency which is operator selectable to generate a secondary frequency between 2.5 kHz and 5.5 kHz. Figure 8 shows a comparison between seabed records made at frequencies of 18 kHz and parametric derived 3.5 kHz.

The vertical resolution of a sub-bottom profiling system is determined by the pulse length of the radiated signal, which in turn is a function of the transmitted primary frequency. The comparatively high primary frequency used by *Parasound* makes it possible to radiate a considerably shorter pulse length than conventional sub-bottom profilers. This together with the ability to generate the low secondary frequency with precise wave periods, results in greatly improved vertical resolution of up to 30 cm, even in deep sea sediments (Fig. 8).

Horizontal resolution is optimized at depths greater than 2000 m by operating *Parasound* in an automatic multi-pulse mode. A pilot pulse is first-transmitted to determine the actual depth of the seabed. The system processor then determines how many pulses will fit into the travel time at that depth and emits a suitable train of pulses. This sequence is then repeated and corrected with the measured depth automatically. This approach greatly enhances the resolution of small-scale features such as sedimentary bed forms (Figs 9 and 10).

In order to achieve optimum penetration and resolution of the sub-bottom layer, account must be taken of differences between the behaviour of sound in water and in sediment. To set and maintain the bottom TVC (Time Volume Control) independently of the water column TVC with changing water depth has always been difficult with earlier equipment. *Parasound* makes use of its relatively high primary frequency to accurately define the true seabed and therefore the start depth for the bottom TVC. The correct setting is then automatically maintained as the depth varies.

A high-definition colour display unit is incorporated within the system to permit monitoring of the operation. The display screen can be split to allow a general overview of the bottom and a high-resolution display of the sediment layers. The display may be frozen and photographed if required (Fig. 12).

A high-precision echogram with depth digitization provides a further record. It is possible to make simultaneous recordings of the bottom profile and sediment layers with annotation of the selected operating parameters.

Conclusions

The Atlas *Hydrosweep* has been proven to offer significantly improved performance levels over other swathe-mapping systems, not only with respect to time saved, but also in the quality of data acquired. The system accuracy is improved despite the wider swathe coverage due to the unique self-calibration process, which also allows depth data to be collected without interrupting other ship routines. Indeed, terrain following navigation capabilities are enhanced.

The Atlas *Parasound* has proved its capability for carrying out high-resolution sediment surveys from hull-mounted transducers with horizontal resolution comparable to that of towed systems, and much better vertical resolution. The narrow beam function inherent in the parametric principle has provided the additional capability of high-precision bathymetric surveys in both deep and shallow water.

References

Berktay, H. O., 1965, Various Papers on Non-linear Acoustics, *J. Sound Vib.*

Blackinton, J. G., Williams, J. F., Hills, D., and Kossalos, J. G., 1986, First Results from a Combination Sidescan and Seafloor Mapping System, *Proc. Offshore Tech. Conf.*, OTC, 4478.

Cloet, R. L. and Edwards, C. R., 1986, The Bathymetric Swathe Sounding System. *The Hydrographic Journal* **40**.

Colladon, J. D., and Sturm, F. K., 1827, The Compression of Liquids (in French). *Ann. Chim. Phys.* Series 2, **36**, Part IV. Speed of Sound in Liquids, 236–257.

de Moustier, C., 1988, State of the Art in Swathe Bathymetry Survey Systems *Int. Hydro. Rev.* **65**, 29–38.

de Moustier, C., and Kleinrock, M. C., 1986, Bathymetric Artifacts in Seabeam Data, *J. Geophys. Res.* **91**, 3407–3434.

Farr, H. K., 1980, Multibeam Bathymetric Sonar Seabeam and Hydrochart, *Marine Geodesy* **4**, 77–93.

Grant, J. A., 1985, Filtering, Digitisation, Logging and Presentation of Depth. *Proc. 3rd Hydro. Symp.* Min of Defence.

Schenke, H. W., 1988, Unpublished Correspondence on Meteor Expedition M6 4, Alfred Wegener Inst. Bremerhaven, West Germany.

Schenke, H. W., and Ulrich, I., 1986, Flachenhafte Kartierung des Meeresbodens, *Geowissenschaffen in unsere Zeit* **4**.

Schreiber, R. and Schenke, H. W., 1989, Atlas Hydrosweep, Efficient Hydrographic Surveying of EEZ with New Multibeam Echosounder Technology for Shallow and Deep Water, *Proceedings EEZ Resources: Technology Assessment Conference*, JOTC Hawaii, U.S.A., Jan. 1989, pp. 3–16 to 3–30.

Schreiber, R., Wentzell, H. F., and Ziese, R., 1988, Efficient Seafloor Mapping and Sub-Bottom Profiling with Atlas Hydrosweep and Atlas Parasound, in *Proceedings of Techno Ocean '88 Conference, Kobe, Japan, Nov. 1988*, pp. 103–110.

Siedler, G. A., Kuhl, A., and Zenk, W., 1987, The Madeira Mode Water, *J. Phys. Oceanogr.* (in press).

Stenborg, E., 1987, Swedish Parallel Sounding Method. *Int. Hyd. Rev.* **64** (1), Monaco.

Tyce, R. C., 1987, *Workshop Report on Hydrosweep Data Postprocessing*, Univ. Rhode Island.

Urick, R. J., 1983, *Principle of Underwater Sound*, McGraw Hill, Inc.

Wentzell, H. F. and Ziese, R., 1988, New Echosounding Methods for Shallow Water and Deep Sea Surveying, *Advances in UW Tech. Vol. 16 Oceanology*, pp. 25–41. Graham & Trotman, London.

GLORIA Image Processing: The State of the Art

R. C. SEARLE*

Institute of Oceanographic Sciences, Deacon Laboratory, Wormley, Godalming, Surrey, GU8 5UB, UK

T. P. LE BAS

Institute of Oceanographic Sciences, Deacon Laboratory, Wormley, Godalming, Surrey, GU8 5UB, UK

N. C. MITCHELL

Institute of Oceanographic Sciences, Deacon Laboratory, Wormley, Godalming, Surrey, GU8 5UB, UK
Department of Earth Sciences, University of Oxford, Parks Road, Oxford OX1 3PR, UK

M. L. SOMERS

Institute of Oceanographic Sciences, Deacon Laboratory, Wormley, Godalming, Surrey, GU8 5UB, UK

L. M. PARSON

Institute of Oceanographic Sciences, Deacon Laboratory, Wormley, Godalming, Surrey, GU8 5UB, UK

and

PH. PATRIAT

Institute de Physique du Globe de Paris, Tour 14–15, 4 place Jussieu, 75230, Paris, France

(Received 27 April, 1989; accepted 1 September, 1989)

Key words: GLORIA, sidescan sonar, image processing.

Abstract. This chapter presents a summary of the image-processing techniques being used at present in the Institute of Oceanographic Sciences Deacon Laboratory's GLORIA long-range sidescan sonar system. It begins with a brief review of the development of GLORIA, and then describes in outline the present shipboard data acquisition, recording and replay system, including simple image-processing techniques that can be used on-board ship. Next, a detailed form of the sonar equation is developed, and this is evaluated factor-by-factor, to demonstrate the effects of beam directivity, refraction and water depth on the form of intensity variation to be expected in the final image. Finally, we discuss recent developments in shore-based image-processing. These include the development of improved radiometric corrections to normalize range-dependent intensity variations, recovery of true backscattering levels and estimation of backscattering coefficients, and combination of GLORIA with other data sets into single, colour digital images. As an example of the last process we show a digital mosaic of sonar data from the Southwest Indian Ridge, coloured as a function of depth derived from Sea Beam data in the same area.

Introduction

The GLORIA long-range sidescan sonar system (Figs 1 and 2) has now been used for about twenty

* Present address: Department of Geological Sciences, University of Durham, Durham, DH3 1LE, UK

years, during which time it has evolved from a "test-bed" instrument to a highly sophisticated and reliable survey tool (Rusby, 1970; Somers, 1970, 1973; Rusby and Somers, 1977; Somers *et al.*, 1978; Laughton, 1981; EEZ-scan 84 Scientific Party, 1988). A major part of this evolution has been concerned with improving the image quality and presentation, and in this chapter we shall briefly outline those developments, and then describe the state of the art at IOSDL in 1988.

The first images had low signal-to-noise ratios, were output on wet-paper recorders, and suffered from severe distortions of both geometry and signal intensity. Slant-range, not horizontal range, was displayed (so the horizontal range scale was nonlinear), no corrections were made for varying tow-speed, and even the average along-track and cross-track scales were different. The recorders used were unable to cope with the >35 dB dynamic range of the signal, so strong signals were saturated and weak ones were lost.

Gradually these defects were remedied, with all improvements to the GLORIA system subsequent to 1971 being ultimately enabled by the introduction then of a linear correlation processor. This involves

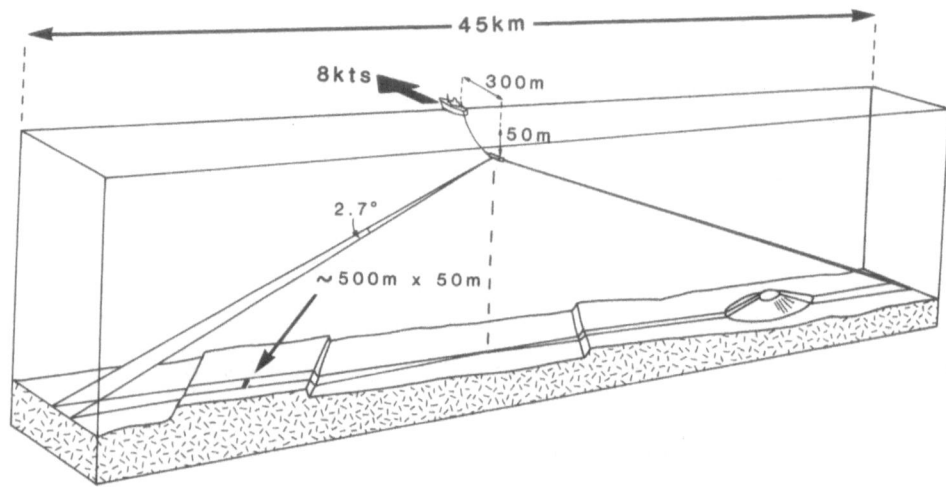

Fig. 1. Schematic diagram of the GLORIA tow-vehicle and its sound swath, with principle dimensions.

the use of a long transmission pulse (up to 4 s) coded with a linear FM sweep. The correlator compresses the energy of the long pulse into a short interval τ that is inversely proportional to the FM bandwidth. In a system with a power-limited trans-

mitter, the receiver signal-to-noise ratio depends upon the pulse energy, while the correlator restores the range resolution. The effect in GLORIA was equivalent to raising the transmitter power of a short pulse system from 10 kW to 4 MW. This enabled

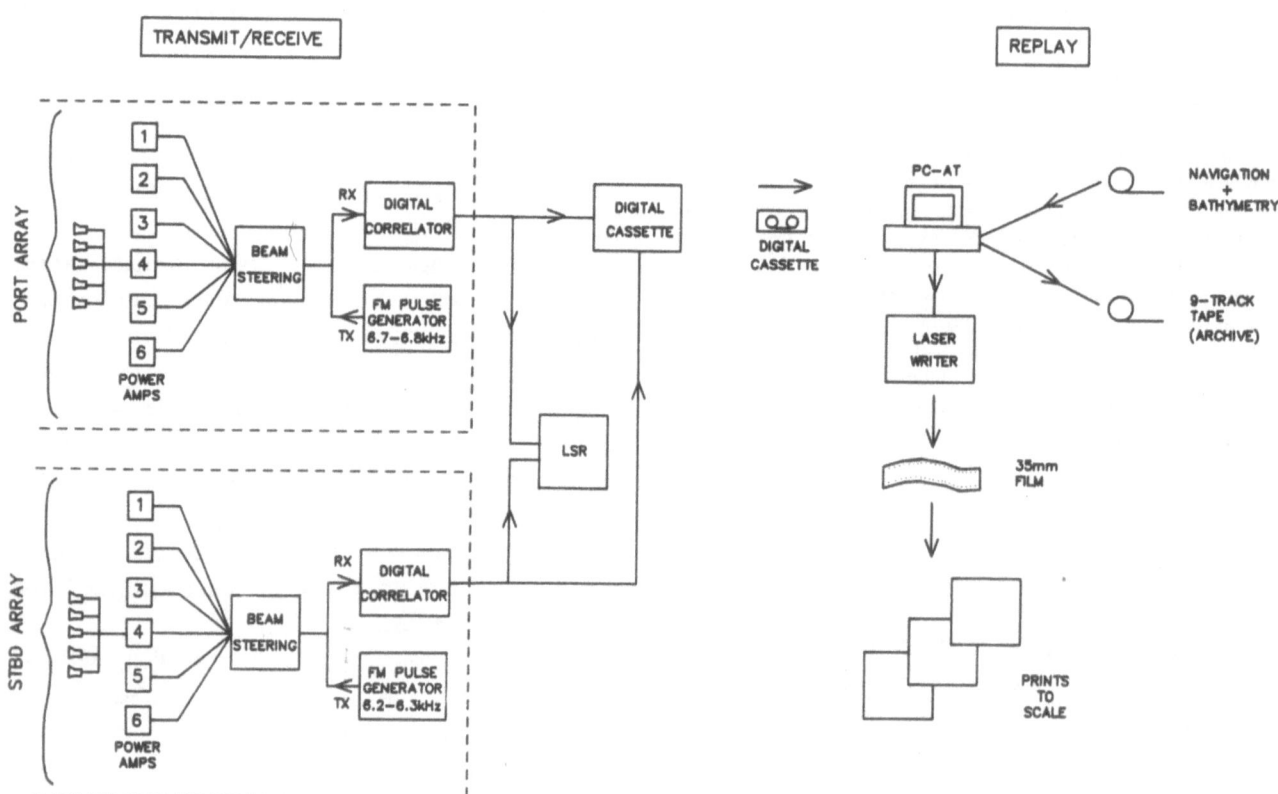

Fig. 2. Simplified block diagram of the present GLORIA shipboard data acquisition and processing system.

adequate S/N ratios to be obtained over long ranges, gave the reduction in vertical array aperture needed to obtain more even illumination of the swath, and thus allowed the introduction of a smaller, more robust vehicle configuration in GLORIA Mark II.

Signals were then replayed from tape through an analogue facsimile recorder to produce good-quality photographic prints. An 'anamorphic camera' was developed to correct these for varying tow-speeds. This applied a variable along-track stretch by rephotographing a moving copy of the original print through a slit on to a moving film, with the ratio of print speed to film speed set by the ship's speed over the ground. At the same time a factor was applied to set the average horizontal range scale equal to the along-track distance scale. To cope with the large dynamic range, data were recorded (on analogue FM tape recorders) in two ways: with automatic gain control (AGC), and so-called 'fixed gain' (FG). In fact time-varied gain (TVG) was applied to both to provide a rough compensation for spreading and attenuation losses. Thereafter, the FG signal was recorded directly, and preserved the true relative intensity of strong signals, but lost weak ones, whereas the AGC channel, in which the recorder gain was automatically reduced for stronger signals, preserved signals of all intensities but on a non-linear, and unrecoverable, intensity scale.

This processing system was introduced in the early 1970s, and within its limitations produced good-quality records throughout that decade. The major limitations were the inability to correct for slant-range distortion and the continued need for AGC recording. There was a further undesirable effect of the AGC, associated with the correlator system. AGC was applied before correlation, and the non-linearity produced in this way caused image distortions, of which the most noticeable was the tendency for a strong echo to suppress the reverberation just before and just after itself, producing erroneous shadows. Even so, in the early 1980s we were beginning to apply a slant range correction to digitized tracings of interpreted features (e.g. Searle and Hey, 1983), and improved tape recording technology, together with the replacement of the earlier analogue signal correlators by digital ones in 1982 and introduction of an improved TVG law on EPROM in 1987, improved the dynamic range of the FG records considerably. In 1980 the analogue signal processing and recording system was replaced by a digital system, opening up a wide range of further image correction, enhancement and processing possibilities. The rest of this chapter will describe our progress in these areas to date.

Current Data Acquisition and Processing Systems

Figure 2 shows a schematic diagram of the present shipboard signal-processing system. The towed sonar vehicle carries two rectangular transducer arrays (one port, one starboard), with their axes directed 20° below the horizontal. Each array consists of 2×30 transducers 0.17 m in diameter and spaced 0.17 m apart, separately wired into six horizontal sections of 5×2 transducers. Each section has independent transmit and receive signals.

Figure 3 shows the computed beam pattern for this arrangement, which produces a main beam $2.7° \times 35°$ at half power points. The horizontal beam pattern has the sidelobe arrangement of any uniformly illuminated line. The effect of these sidelobes is to broaden the beam somewhat, and this can be seen especially on distant discrete targets. The vertical beam pattern is the non-ideal compromise between several conflicting demands:

- for mechanical engineering reasons a minimum vehicle diameter was needed;
- the peak of the main vertical lobe should ideally be about 5°–10° below the horizontal;
- the sensitivity should reduce uniformly to a low value at the nadir;
- and the sensitivity should reduce sharply above the horizontal.

The low sensitivity in the nadir is required to reduce the occurrence of second and even third bottom echoes over flat terrain, which were so unwelcome a feature of the GLORIA I records. In fact the pattern has a small null 20° or so off the nadir, and second bottom echoes are still occasionally a problem with GLORIA II.

The transmitted signal is a 100 Hz linear FM pulse, usually 2-s long repeated every 30 s, although other pulse repetition periods (PRPs) and pulse lengths are available. Separate carrier frequencies of 6.2875 kHz (starboard) and 6.7625 kHz (port) are used to reduce cross-talk to a negligible level. Normally only three sections per side are used for transmission, so the transmit beam is 5° wide, allowing

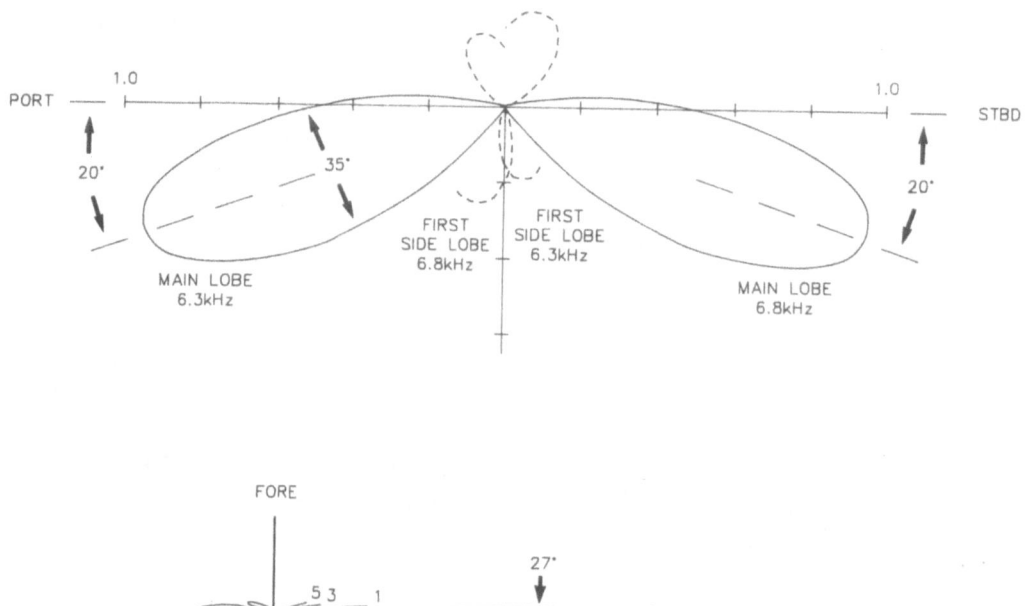

Fig. 3. Power directivity diagrams for the GLORIA array. Top: vertical section normal to axis of vehicle. Bottom: horizontal section; right half shows pattern of whole array (six sections), left half that of a single section to the same scale. Main lobe and first five side lobes are numbered 0 to 5; each side lobe is one of a symmetric pair, of which only one (alternating fore and aft) is shown for clarity.

good insonification even in the presence of vehicle yaw. All sections are used for reception, and after preamplification have TVG applied using the law

$$T = -30 \log R - 8.2 \times 10^{-4} R \qquad (1)$$

to match spreading losses and attenuation in the water column (see equations (4) and (7), next section), where T is gain in dB and R is slant-range in m. This assumes the attenuation coefficient α is 0.82 dB km^{-1}. The signals from the separate sections are then phased to steer the beam along the normal to the long-period mean heading of the vehicle. This electronic beam steering is effective for yaw up to about $\pm 5°$. Higher yaw may be experienced in sea states of roughly 6 and over, and can produce muted lines ('dropouts') in the records.

After beam formation, the port and starboard signals are correlated and digitized into 496 samples per side. (In fact the digitization rate is 500 samples per side, but for technical reasons the last 4 samples are discarded, causing no real loss since their information content is heavily eclipsed by the next transmission pulse). This yields a range resolution of 45 m

for a 30-s PRP. At 8 knots (4 ms^{-1}) the near-ship along-track resolution is 120 m for the same PRP, although this resolution is degraded by the finite beam width to about 900 m at maximum range. The data are then recorded on digital tape cartridges with a sampling rate set to ensure 992 pixels per scan (496 per side) and simultaneously displayed on a paper line scan recorder monitor. Thus for a 30 s PRP the pixel size is 60 ms or 45 m. The remaining 32 bytes of the 1024-byte record are used as headers to record other pertinent information (see Table I). The output from the correlators is a 12-bit signal, but prior to recording this is compressed to 8 bits using the "Bell (MU) 255 law"

$$y = C \ln(1 + 255x/4096), \qquad (2)$$

where x and y are the input and output signals, respectively, and C is a constant. Each cartridge normally holds two data files (or "passes" in our jargon), and for a 30-s PRP, each pass will contain six hours' worth of data in 720 scans.

Individual scans to the LSR monitor are replicated so that the record is approximately isometric for a

TABLE I

Recording format used for GLORIA data

Byte number	Word in IBM–AT	Contents
1	1 Low	Pass number (binary, high byte)
2	1 High	Pass number (binary, low byte)
3	2L	Scan number (binary, high byte)
4 ·	2H	Scan number (binary, low byte)
5	3L	Flag for hour mark
6	3H	Slant-range correction code
7	4L	Binary zero
8	4H	Pulse repetition period, seconds (ASCII)
9–11	5L–6L	Vehicle heading, degrees (3 ASCII characters)
12–13	6H–7L	Year (2 ASCII characters)
14–15	7H–8L	Hexadecimal FF (edge mark)
16	8H	Sonar sample 1, far range port (binary)
:	:	:
512	256L	Sonar sample 496, near range port
513	256H	Sonar sample 497, near range starboard
:	:	:
1009	505L	Sonar sample 992, far range starboard
1010–1011	505H–506L	FF (edge mark)
1012–1014	506H–507H	Julian day (3 ASCII characters)
1015–1016	508L–508H	Hours (2 ASCII characters)
1017–1018	509L–509H	Minutes (2 ASCII characters)
1019–1020	510L–510H	Seconds (2 ASCII characters)
1021–1022	511L–511H	Checksum
1023–1024	512L–512H	Unused

ship speed of about 8 knots, and a rough slant-range correction is applied via a look-up table using depths entered manually by the watchkeeper.

At intervals throughout a cruise (usually every 24 h), the digital data are processed further. The reason for using cartridges will now be clear, since with a data rate of 3 Mbyte per 24 hours, a 2400-foot 9-track tape would hold over four days' worth of data, whereas each track of a cartridge holds a convenient six hours' worth, leading to a much more convenient shipboard replay schedule. The cartridges are downloaded to disc on an IBM AT personal computer, and also transferred to standard 9-track magnetic tape as an archive medium. The cartridges are wiped clean and reused once the 9-track tapes are safely duplicated and at separate locations ashore, but enough are available on the ship to hold 100 days' worth of data.

The AT is then used to apply the anamorphic and slant-range corrections in a precise way, using navigation and depth data supplied (at present by tape transfer) from the ship's normal computer logging system. The anamorphic correction is achieved by computing the number of times an individual scan must be replicated to achieve the desired "stretch" (since this is in general not an integer, slight over- and under-corrections alternate). Slant-range correction is achieved by computing the horizontal range for a given slant-range, assuming a plane horizontal sea floor whose depth is that below the ship, and moving the contents of the recorded pixel to the appropriate new pixel. (Note that this implies some replication of pixels, because the real time data and the slant-range corrected data have the same number of pixels, but the direct water path has been removed from the latter).

We can also carry out further processing at this stage: for example, filtering to remove line drop-outs using the high-pass/low-pass method of Chavez (1986), application of radiometric corrections (along the lines to be outlined in a later section of this paper) to produce more uniform 'illumination' across the image, and reversal of image polarity (white on black or black on white presentation).

The processed files are then written to 35 mm film on a purpose-built laser film-writer. The film is developed on board, and photographic prints are made to the desired survey scale (usually between 1:500 000 and 1:250 000; 1:375 000 has proved to be a particularly popular and appropriate scale). If a 24-hr data batch ends in the morning, the complete processing cycle, including production of photographic prints, can be completed by the same evening, so that no data are delayed longer than 36 hr from recording to arrival on the plotting table.

Further processing can be carried out ashore. At present this is done at IOSDL, Wormley, using the United States Geological Survey's MIPS (Mini Image Processing System) software (Chavez, 1986) run on a PDP 11/34 computer with Grinnell monitor and interface (In fact the computer was upgraded to a Micro Vax II while this paper was in press). However, processing is not hardware-dependent, and one of us (NCM) has done a considerable amount on a Sun workstation at Oxford University.

Acoustic Considerations

Before describing in detail our progress in shore-based processing, it is instructive to consider quantitatively the effects on the GLORIA images of both the sonar system itself and the propagation of sound through the sea.

We use spherical coordinates where r is the range from the sonar array, ϕ is the angle in the vertical plane normal to the axis of the array, and θ is the angle in the horizontal plane through the array. The acoustic intensity I is the incident power P per unit area A and is related to r.m.s. pressure variations p by

$$I = P/A = p^2/\rho c, \qquad (3)$$

where ρ is the density and c the sound velocity of seawater. Since the GLORIA transducers are linear in pressure, their output voltage should be proportional to the square root of the acoustic intensity.

At some distance r from a source of sound of power P_0, directivity factor D and directivity pattern $b(\theta, \phi)$, the incident sound intensity will be

$$I_i(r, \theta, \phi) = P_0 D \cdot b \cdot \exp(-\alpha r)/4r^2. \qquad (4)$$

The dimensionless backscattering coefficient S_b is defined as

$$S_b = P_b/I \cdot A, \qquad (5)$$

where P_b is the backscattered power and A is the effective insonified area; the latter can be approximated by

$$A = (R\theta_b \cdot c\tau \cdot \sec \beta)/2, \qquad (6)$$

where R is the range of the backscattering point, θ_b is the -3 dB horizontal beam width, τ is the width of the function produced by passing the transmitted FM pulse through the receiver correlator, and $\beta(R, \phi)$ is the grazing angle of the ray incident on the bottom (in general not equal to ϕ because of refraction in the water column).

Combining (4), (5), and (6), the backscattered power at the seabed is

$$P_b = P_0 D \cdot b \cdot \exp(-\alpha R) \cdot S_b \theta_b \cdot c\tau \sec \beta/8\pi R.$$

Allowing for further attenuation and spherical spreading on the return path, the intensity incident on the receiver array is

$$I_b = P_0 D \cdot b \cdot \exp(-2\alpha R) \cdot S_b \theta_b \cdot c\tau \sec \beta/32\pi^2 R^3,$$

and taking account of the receiver directivity, the received acoustic power is

$$P_r = P_0 D^2 b^2 \cdot \exp(-2\alpha R) \cdot S_b \theta_b \cdot c\tau \sec \beta/32\pi R^3.$$

This produces a voltage $V \sim P_r^{\frac{1}{2}}$, and applying TVG ($T(R)$, Equation 1) and other constant receiver gains (G), we have

$$V = (P_0 S_b \theta_b \cdot c\tau \sec \beta)^{\frac{1}{2}} DbTG$$
$$\cdot \exp(-\alpha R)/4 \cdot (2\pi)^{\frac{1}{2}} R^3, \qquad (7)$$

In evaluating expression (7) we must remember that for a given horizontal range both the grazing angle β and the initial ray angle at source (and hence D) will depend on R, the slant range measured along the acoustic ray. Moreover, the rays in general will follow curved lines because sound velocity varies with depth, giving rise to refraction. Moreover, this refraction may cause rays to converge or diverge more than expected for spherical spreading in a constant velocity ocean, thus changing the incident sound intensity. We define

$$F = I_i(\text{aspherical})/I_i(\text{spherical}) \qquad (8)$$

as the "refraction factor". So finally we have

$$V = (P_0 S_b \theta_b \cdot c\tau \cdot \sec \beta)^{\frac{1}{2}} DbTGF$$
$$\cdot \exp(-\alpha R)/4 \cdot (2\pi)^{\frac{1}{2}} R^3$$

or, explicitly including T (Equation 1):

$$V = (P_0 S_b \theta_b \cdot c\tau \cdot \sec \beta)^{\frac{1}{2}} DbGF/4 \cdot (2\pi)^{\frac{1}{2}}. \qquad (9)$$

In the following sections we illustrate some of the features of this equation.

RAY TRACING

In order to examine the effects represented in Equation (9), we need to evaluate them for appropriate values of the slant range R. We did this by using a standard ray-tracing program to compute where rays emitted by GLORIA, at a typical tow depth, would strike the sea floor. Figure 4a, b shows two such plots for typical operating conditions and two different water depths. On these plots, which have equal vertical and horizontal scales, the ray bending due to refraction can clearly be seen. Note too that the rays are tending to converge or bunch up near the upper edge of the beam. The shallowest ray shown here is emitted at $10°$ above the horizontal; rays taking off at higher angles are reflected downward by the sea surface even more steeply (or scattered by it in rough weather), so the effective top surface of the beam is roughly limited by the ray shown. Thus two important features are clear from the figure: first, the maximum range is limited by refraction (i.e. there is the well-known—and militarily important—shadow zone between the top of the beam and the sea surface), and this maximum range increases with increasing water depth; secondly, rays can become focussed near the top edge of the beam, giving rise to rapidly varying sound intensity, as we shall see below.

BEAM DIRECTIVITY

Figure 4c, d shows the directivity pattern $b(\phi)$ in the vertical plane mapped on to the horizontal plane of the sea floor, by evaluating it at the appropriate value of ϕ for each ray and plotting it against the horizontal range where that ray strikes the sea floor. This is done for the same two water depths and the same velocity profile as illustrated in Fig. 4a, b. The important features to note are that

(1) the maximum of the function (centre of main beam) occurs closer to the ship, and the response falls off more rapidly with increasing range, in shallower water; and

(2) the minimum (the null between the main beam and first sidelobe) occurs closer to the ship and is narrower in shallow water than in deep.

The loss of signal in this null is significant, especially in deep water and, as we shall describe later, we have put considerable effort into ameliorating this effect in our image processing.

REFRACTION FACTOR

To evaluate the refraction factor F (8), we note that the incident intensity at the sea floor is inversely proportional to the spacing of rays there, and compare this with the spacing in a uniform velocity ocean (i.e. straight rays from the source). The resulting factor is illustrated in Fig. 4e, f. In the shallow water case (Fig. 4e) this factor increases gradually with increasing range, and beyond about 10 km climbs steeply to a value of about 2 at maximum range. The cause is the tight bunching of rays near the upper edge of the beam (Fig. 4b). The effect is critically dependent on the slope of the thermocline, and Fig. 5 shows the effect of increasing this slope in the region of the sound source, while keeping the other parameters the same. This produces an enhanced response at far range, which shows up on the records as a line or band of increased intensity at the edge of the sonar coverage (Fig. 6). It is usually most severe in shallow tropical or subtropical waters, where the restricted range and strong refraction due to a well-developed thermocline combine to facilitate it. Since the causative refraction mostly takes place near the sea surface, local variations in acoustic velocity there can modify the effect; in particular, the presence of internal waves at density interfaces in the near-surface waters can strongly modulate the focussing effect, leading to the appearance of speckling or even coherent fringes in the high-intensity band. By contrast, in deep water the refraction factor remains close to unity throughout the range, since in this case the strongly focussed upper part of the beam does not intersect the sea floor within the range of the sonar (Fig. 4a).

TOTAL EFFECT

The spherical spreading ($1/R^2$), incidence area ($\sec \theta/R$), and attenuation terms all vary monotonically and show no particular points of interest; except for the $\sec \theta$ term they should be offset by the system TVG. In Fig. 4g, h, we show the combined

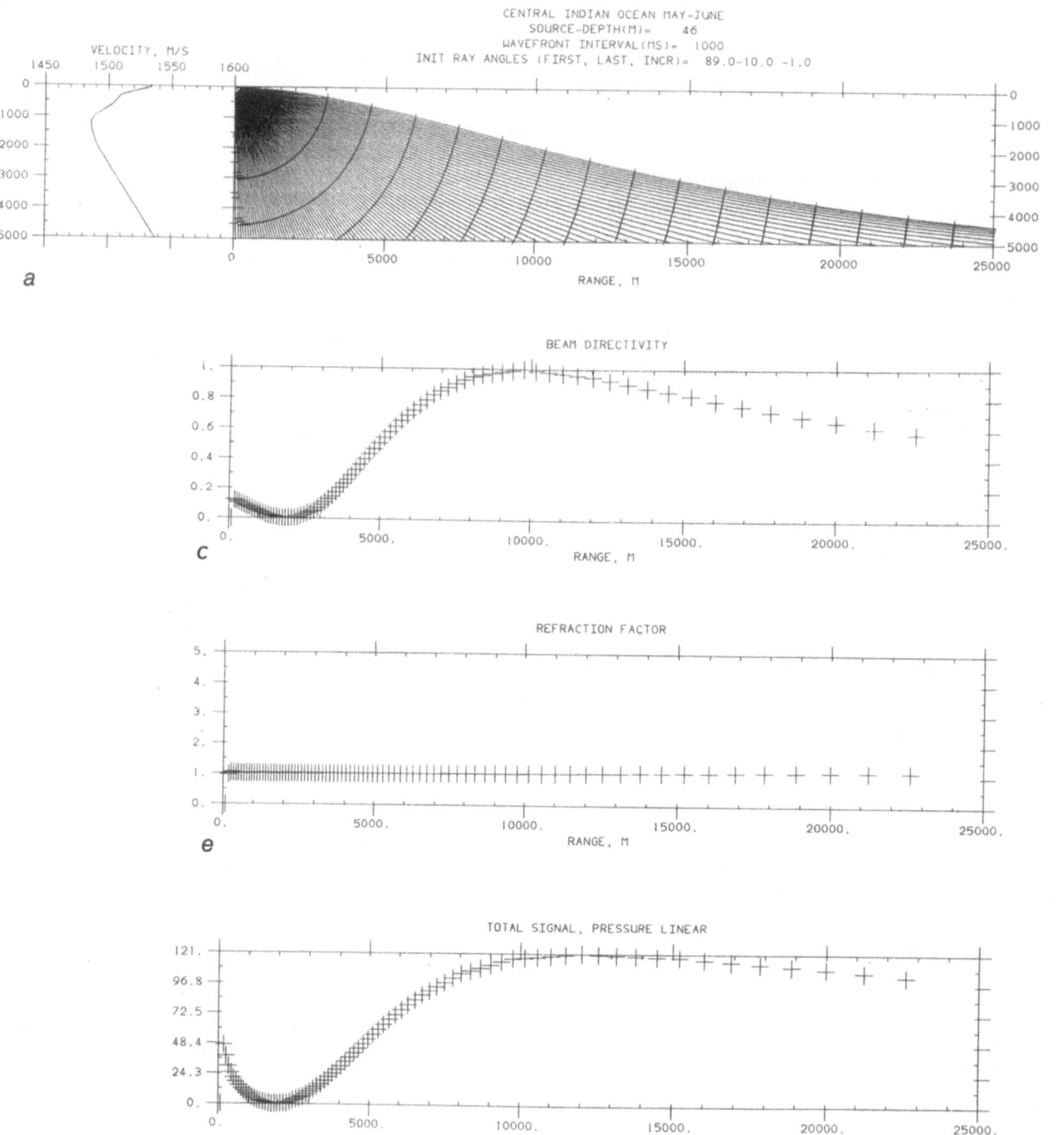

Fig. 4.

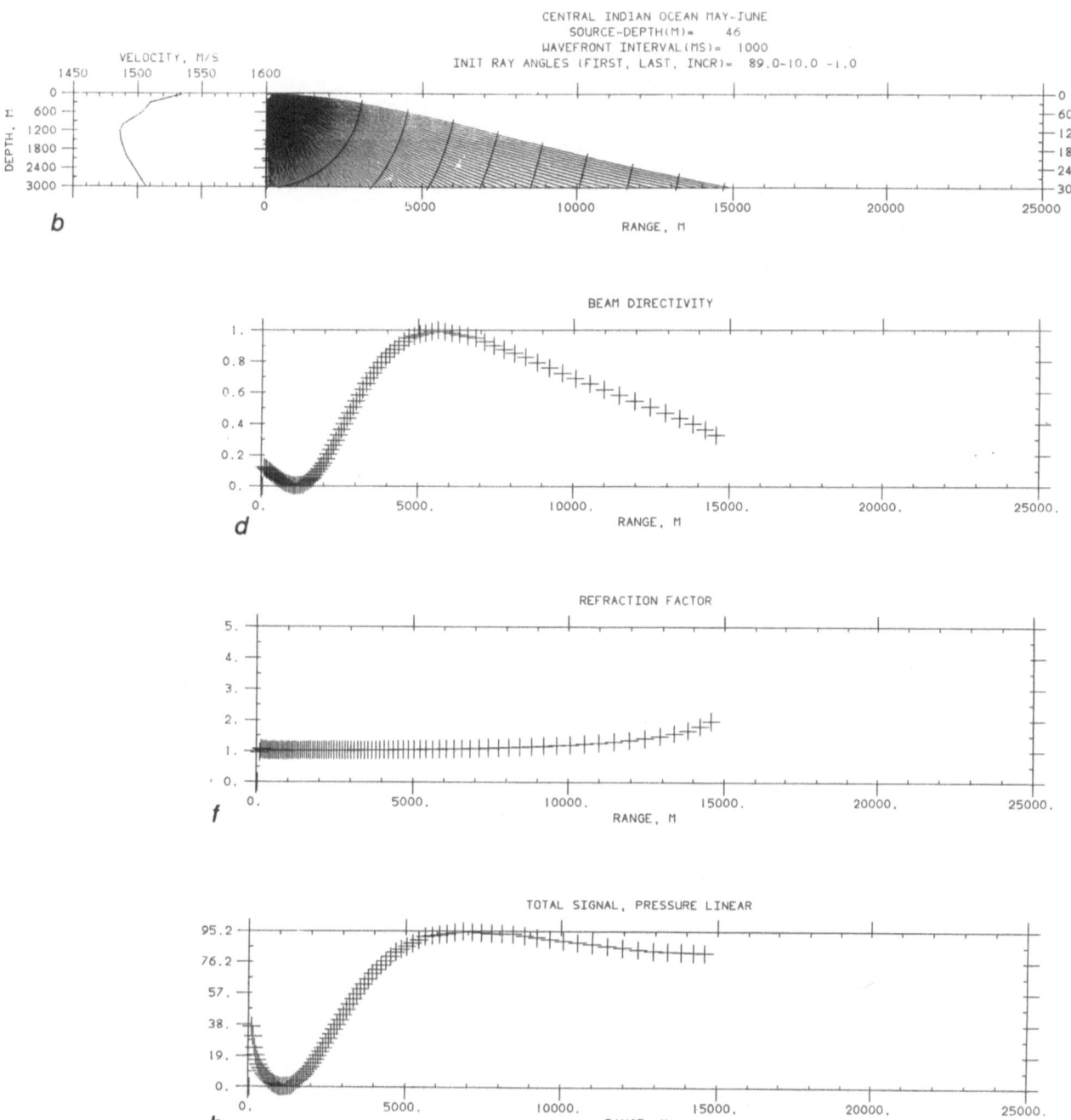

Fig. 4. Computed ray diagrams and effects of various instrument and propagation effects on GLORIA image intensity for typical operating conditions. Left-hand column computed for sea floor at 5000 m, right-hand column for 3000 m. Horizontal scale is the same throughout. (a, b) ray paths. The assumed velocity/depth function is shown on the left (based on data from the Central Indian Ocean in May–June). Right-hand part of plot shows rays with take-off angles 1° apart between −10° (above horizontal) to 89° (near nadir). Vertical and horizontal scales are equal. Arcs show wavefront at 1 s intervals. Sound source at 46 m. (c, d) the beam directivity pattern (b). (e, f) the refraction factor $F = I_i(\text{aspherical})/I_i(\text{spherical})$. (g, h) the product of all effects (text Equation 9) assuming a constant backscattering coefficient $S_b = 1$.

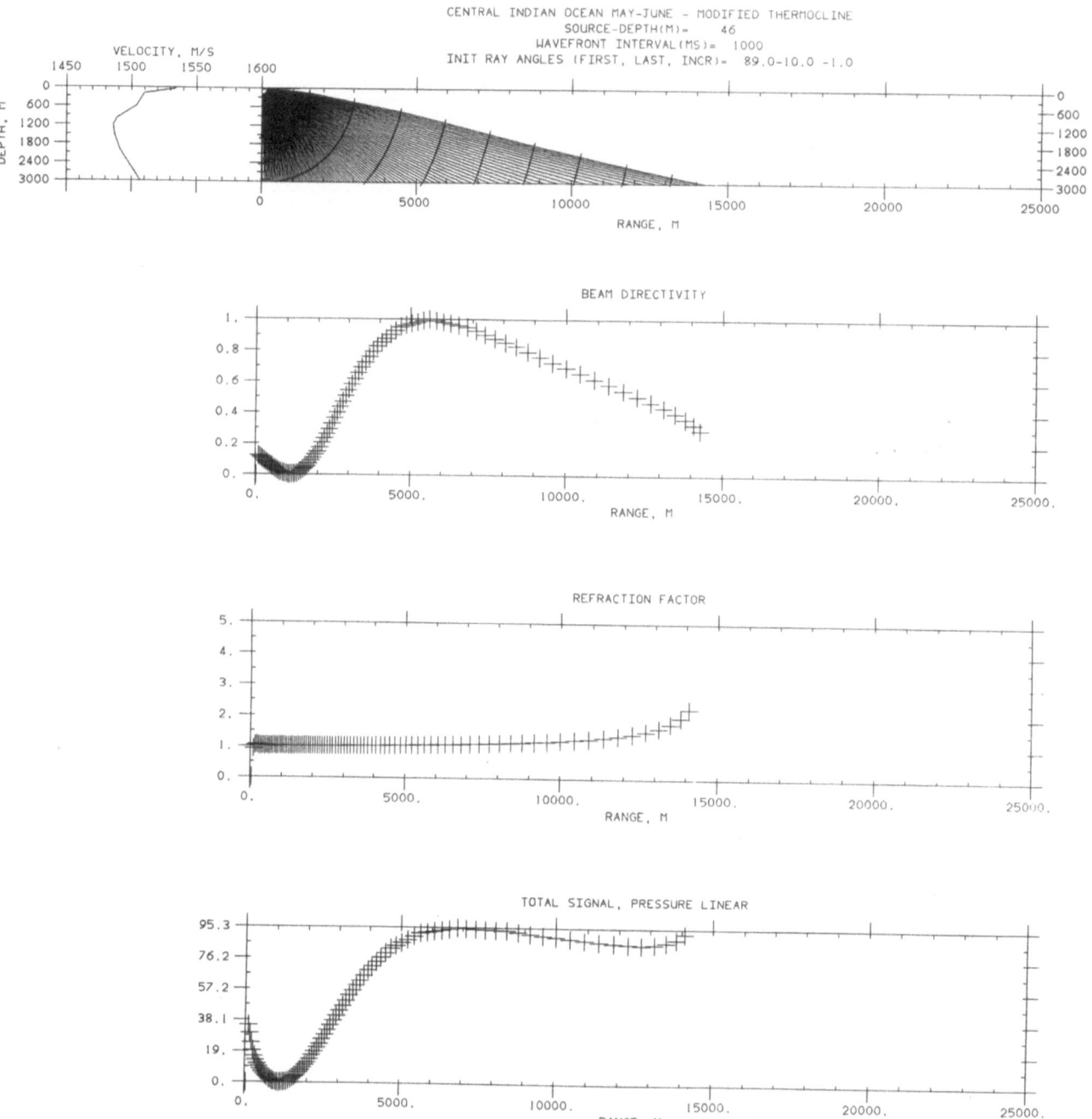

Fig. 5. As Fig. 4, but using a slightly modified velocity profile in which there is a steeper thermocline in the vicinity of the sound source. Note that this leads to increased response at far range due to enhanced refraction.

effect of all the factors. This is the same as equation (9) for a constant backscattering coefficient $S_b = 1$. The variation of intensity with range is dominated by the effect of the beam directivity, especially the null between the main beam and the first sidelobe. Other features, such as the gradual fall-off in gain at mid-

to-far range, and the refraction caustic, are also preserved.

However, although these features are qualitatively similar to those observed in actual signals, we find the quantitative match is variable. In particular, the response at far range often falls off somewhat faster

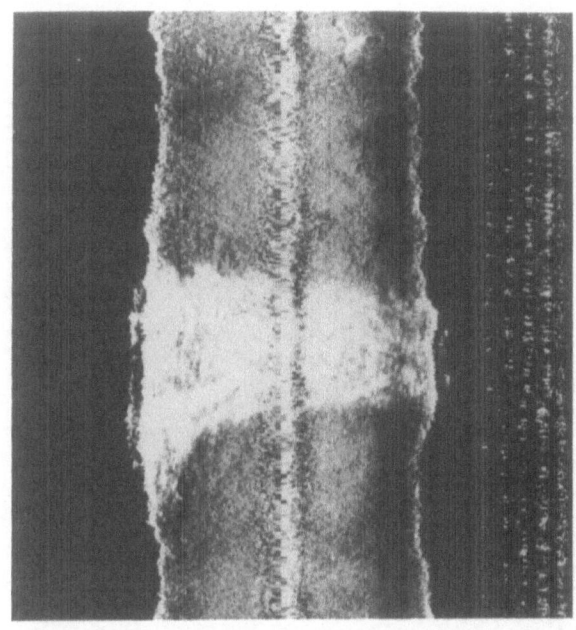

0   KM   10

Fig. 6. Examples of the caustic produced by focussing of the outer part of the beam as a result of near-surface refraction (narrow white bands at the edges of the data swath). Black areas beyond the caustic lay above the sonar beam so were not imaged. Sonograph from the north flank of Saya de Malha Bank, Indian Ocean, near 9° S, 62° E, water depth about 2000–2500 m. The broad white band crossing the swath arises from a debris flow down the margin of the bank. Parallel, broken white bands along right edge are artefacts.

than Fig. 4 predicts, and the first null is somewhat narrower. The former may be largely a result of using an oversimplified backscattering law in the calculations (we know for example that the fall-off in apparent system response is different for different sea floor lithologies), while the latter may largely be due to poor modelling of the near-nadir beamform.

Image Processing

In this section we describe three areas in which we have made recent progress. These are (1) development of a radiometric or 'shading' correction to further flatten the response curve shown in Fig. 4g, h; (2) implementation of that correction to enable true levels of backscattered energy and values of the backscattering coefficient to be estimated; and (3) digital mosaicking and combination of GLORIA and Sea Beam (multibeam echosounder) data into single images for enhanced presentation of the data.

SHADING CORRECTION

Our first attempt at a shading correction was carried out on the NERC I²S image-processor then installed in Swindon in 1984. The method we adopted was to find an empirical function to approximate the instrument/propagation response (Equation 9). We call this the *standard profile*. It would seem best to calculate the standard profile from Equation (9) itself; however, even though all the appropriate parameters are supposedly known, experience with the similar SeaMARC II instrument (Reed, 1987) has shown that in practice such computed profiles do not accurately reflect actual instrument response. We intend to try this approach in the near future, but for the time being we present the results of the more empirical approach.

The empirical function we used first was a smoothed average of several data scans over a featureless abyssal plain (Searle and Kidd, 1984; Searle and Hunter, 1986). By dividing each pixel of an image by the value of the corresponding (same range) pixel in our standard profile, we effectively removed the major, cross-track instrument response. This was very effective in enhancing the image, giving it more uniform contrast, revealing previously unseen information in the weak near- and far-range areas, and enabling better comparisons between features at different ranges. However, this method is not ideal, since the standard profile also incorporates the effects of possible variations in backscattering as a function of range or incidence angle. Since the exact nature of these variations will probably be a function of the seabed lithology, different standard profiles might be needed for different areas. Moreover, since the standard profile now contains (and this shading method removes) this information, potentially important geological information is being lost.

Meanwhile the USGS followed a slightly different approach (Chavez, 1986). Instead of taking their standard profile from a known featureless area, they use the along-track average of a single 6-hour data pass (implemented by subroutine SHAD2 in the MIPS software). This works well in areas of relatively low relief, and is convenient in that it allows for automatic shading corrections, taking into account the variations in average backscattering between different terrains. It still suffers from the disadvantage that knowledge of these variations is

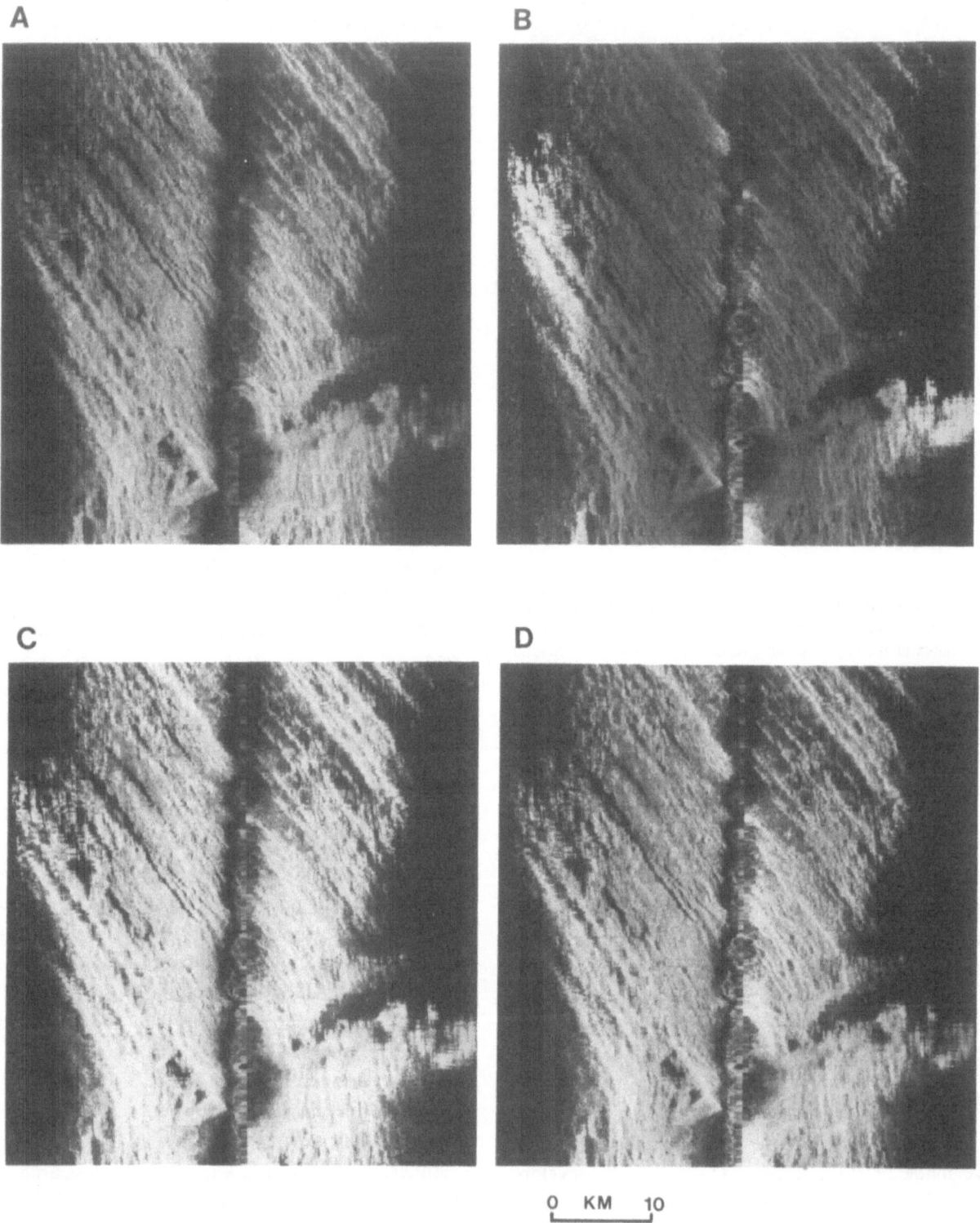

Fig. 7. The results of different forms of 'shading' correction on an image crossing the Rodriguez Triple Junction near 25.5° S, 70° E. Closely spaced lineaments are fault scarps and volcanic ridges of the tectonic spreading fabric; cross-cutting feature is a minor fracture zone. (a) 'raw' image after slant-range and anamorphic corrections; (b), shaded using the along-track average profile (SHAD2): (c), shaded using the along-track maximum profile (SHADX); (d), shaded using the function shown in Fig. 8. Note that *d* particularly produces a more uniform 'illumination', and enhances features at very near and at far range. The possible improvements are usually even greater over well-sedimented areas (see, e.g. Searle and Hunter, 1986).

lost, but it also has a more serious defect in rugged areas. Where there is strong sea floor relief, the maximum range of the sonar beam will vary (Fig. 4a, b). If the pass is now averaged along track, the averages for the outermost pixels will include both real data where the swath is wide, and zeros from areas of shadow where it is narrow. The standard profile will thus be too low at its margins, and the corresponding parts of the corrected image will be too bright (Fig. 7a, b).

To overcome this defect, we construct the standard profile not from the along-track average of all pixels at a given range, but from the maximum occurring along the whole pass for that range (subroutine SHADX). The rationale is that in areas of strong backscattering (which also tend to be the ones with high relief) there is a strong chance that the maximum signal received from a relatively small area will be close to the maximum value that could be received from that area (i.e. the area will contain at least one optimum scatterer), and this will accurately convey the instrument response. Fig. 7c shows the result of applying shading in this way, which is clearly better in this case than the averaging version.

However, even this version of the correction suffers from a dependence on the local geology, and will in general vary from place to place. Therefore, we now employ a user-defined version of the standard profile, which is constructed using the SHAD2 and SHADX profiles as a guide (Figs 7d, 8), This allows the user to tailor the correction to his requirements, including the use of a fixed standard profile over a whole survey area or beyond. In the near future we intend to try a standard profile based entirely on the response computed from Equation (9).

BACKSCATTERING DETERMINATIONS

The backscattering coefficient S_b (Equation 5) is usually quoted in its logarithmic form, $10 \log S_b$ (dB). A large number of acoustic experiments have been carried out on the sea floor to determine the dependence of S_b on angle of incidence, bottom type (roughness and acoustic impedance) and acoustic frequency (for examples see Urick, 1975; McKinney and Anderson, 1964). These typically show a decrease of 20–40 dB with decreasing grazing angle, and a variation of 20 to 25 dB between different bottom types.

In view of the obvious benefits to be expected from direct comparison with experimental results such as these, we have derived a simple model for estimating the backscattering strength from GLORIA data and can display the value of S_b as an image. This allows for quantitative estimates of the likely causes of backscatter observed on sidescan images.

We compute S_b from the logged signal voltages V, using a simple model (equation 9) that assumes only spherical spreading, and approximate values for GLORIA's voltage gains G. Our estimates of the gains and other parameters are probably accurate to within only about 10 dB at present, so our discussion concerns the variation in S_b, not its absolute value. The beam pattern calculation is poor near the nadir (ranges less than 3 km), so the computed values of S_b in that range are unreliable. So may be those at far range, where the effects of refraction have been ignored.

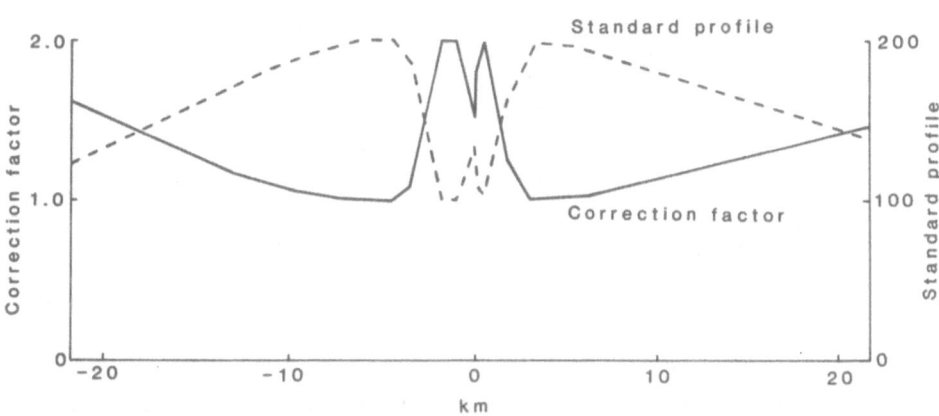

Fig. 8. 'Standard profile', and its inverse the correction factor, used to produce the shaded image shown in Fig. 7d.

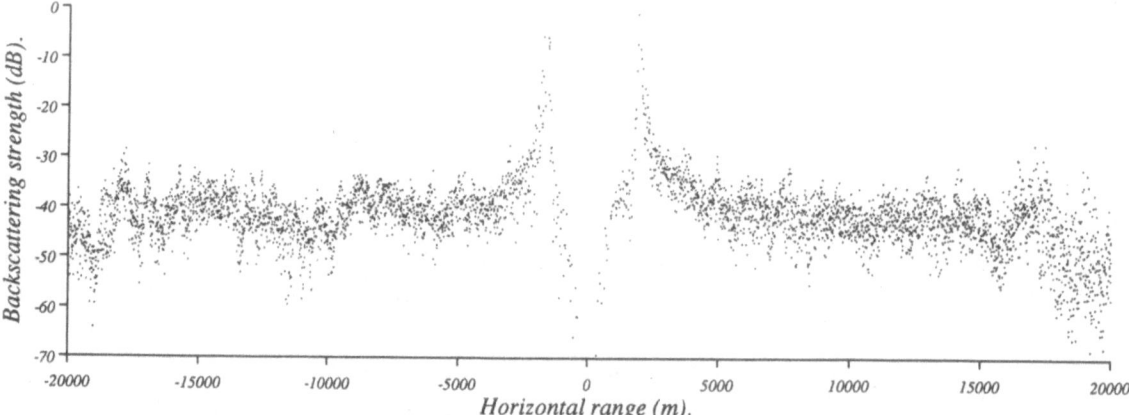

Fig. 9. Profile of inferred backscattering coefficient S_b, against horizontal range, from a sonograph over very thinly sedimented basaltic sea floor (similar to that of Fig. 7) on the crest of the Southeast Indian Ridge near 26° S, 70° E. The scan direction was parallel to the principal topographic fabric. Points from six adjacent scans have been stacked. Absolute values are only reliable to about ± 10 dB because of poorly known instrument gains and propagation parameters, but relative levels should be approximately correct except within 3 km of the nadir, where the beam directivity is poorly modelled.

Figure 9 shows inferred values of S_b plotted against horizontal range. The figure shows the stacked determinations from six adjacent scans parallel to the topographic grain over the crest of the medium spreading rate Southeast Indian Ridge near 26° S, 70° E. This is an area of well-lineated seafloor spreading (abyssal hill) fabric, with only a few metres or less of sediment covering young basaltic lavas. There is a small decrease in S_b with range, consistent with the decrease in average grazing angle. The ~ 10 dB scatter in S_b at a given range is to be expected from changes in bottom roughness and slope. Large decreases (e.g. near -10 km) are probably due to acoustic shadowing by intervening objects.

Figure 10 shows a similar display of S_b from the same area as Fig. 9, but for a pass in which the scans were normal to the lineated topographic fabric. Here S_b displays much greater scatter (> 20 dB) for a given range because of the greater variation in grazing angle for this scan direction. Very low values due to shadowing are also more common, and become more frequent with increasing range.

Despite the shortcomings of this simple model, we feel that reasonable values of S_b can be calculated from a fair proportion of the data. For more accurate estimates, we need to take account of refraction and the effects of real bottom topography. We are now progressing in that direction, estimating the

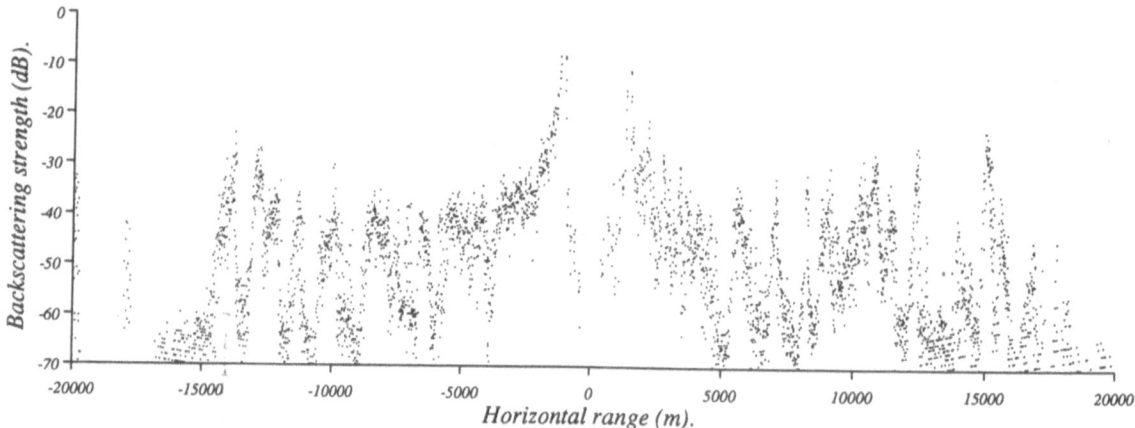

Fig. 10. Profile of inferred backscattering coefficient S_b, against horizontal range, from a sonograph over very thinly sedimented basaltic sea floor on the crest of the Southeast Indian Ridge. As for Fig. 9, except that here the look direction was normal to the principal topographic fabric. Alternating peaks and troughs in S_b are due to reflections and shadows from the sea floor ridges.

refraction effect as outlined previously, and using Sea Beam data to make precise estimates of grazing angle. The results of these investigations will be reported elsewhere.

MOSAICKING AND COMBINATION WITH SEA BEAM

Assuming that the system and propagation parameters are known, the backscattered acoustic intensity recorded in each pixel by GLORIA can be regarded as a function of two variables: the backscattering coefficient S_b (itself a function of the seabed roughness and acoustic impedance), which is linked to surface geology, and the angle of insonification, which depends on range and the slope of the sea bottom. Thus two pixels could contain the same intensity (and be interpreted as similar geological features) even though they actually represent different lithological and topographic entities. It is therefore important for a proper interpretation to consult other sources of data, particularly bathymetry, to aid geological identification and correlation. This can be a slow operation, and the initial absence of the supplementary information reduces the ease of understanding of a mosaic of raw GLORIA images.

The Sea Beam multi-narrow-beam echosounder system (Renard and Allenou, 1979) produces high precision bathymetric maps of the sea floor. Mixing of Sea Beam and GLORIA data in a single digital image can therefore complement the two data sets and aid interpretation. We have used this technique in an area having 100% coverage of GLORIA data and over 80% coverage of Sea Beam data (the lack of full coverage is due to the much smaller (~ 2.5 km) swath width of the Sea Beam system, combined with real time navigational uncertainties of a similar size, and the finite time available for surveys). The area chosen is a section of the Southwest Indian Ridge near 27.5° S, 66° E.

The first step is to produce digital mosaics of each data set. Creating the GLORIA mosaic requires every track which crosses the area to be digitally processed (slant-range, anamorphic and shading corrections, and filtering to remove line drop-outs), producing separate images of each track segment. These are then mosaicked by digitally adding them into a master image in their correct geographical locations, making a final mosaic covering the whole area (Fig. 11). Where segments overlap, the operator can view both images in different colours, and decide

to use parts of one or the other or an average of the two. Our state of knowledge is too primitive at this stage to consider a computed "cut and paste" algorithm, and the difficulties of developing one would be compounded by the fact that overlap areas are in general insonified from widely different directions. However, we have in mind in the longer term an algorithm based on contouring the cross-correlation between small-windowed areas of the overlaid images and looking for a ridge in the contours to determine the best match.

The Sea Beam data have to be handled slightly differently. The input consists of numerical integer depths and the corresponding ranges from ship. Thus for each depth value a location has to be calculated from the position of the ship and the range value (which is perpendicular to the ship's track). The individual point soundings are then placed on the nearest points of a 50-m grid. The results are stored in a large two-dimensional array held in dynamic memory before writing the array to a disk file when all the input data have been read. This 'image' is then transferred to the image-processor for combination with the GLORIA image.

It is more convenient for us to grid the Sea Beam data on the IOSDL mainframe IBM computer rather than the MIPS one, but memory restrictions on the former mean that the Sea Beam data have initially to be assigned to a 50-m grid—a slightly larger spacing than the 45-m optimum for GLORIA gridding. The next operation therefore is to interpolate the Sea Beam data on to a 45-m grid, which can be done by simple linear interpolation without loss of real data. The same interpolation can also make up for the lack of full Sea Beam coverage by filling in gaps where necessary. We now have two images of the same area with the same pixel size for combination, both with 100% coverage.

GLORIA images are displayed as black and white images usually with 256 density levels (or 'grey levels') requiring 8 bits. The Sea Beam images, however, contain integer values between 0 and 15000 m, requiring at least 14 bits. Most MIPS programs are designed to work with 8-bit data and therefore the Sea Beam data need compression. A new scale is given to the data where 0 represents the shallowest depth (in this case 2000 m) and 255 the deepest data (5500 m). Thus vertical resolution of the data was reduced in this case from 1 m precision to about

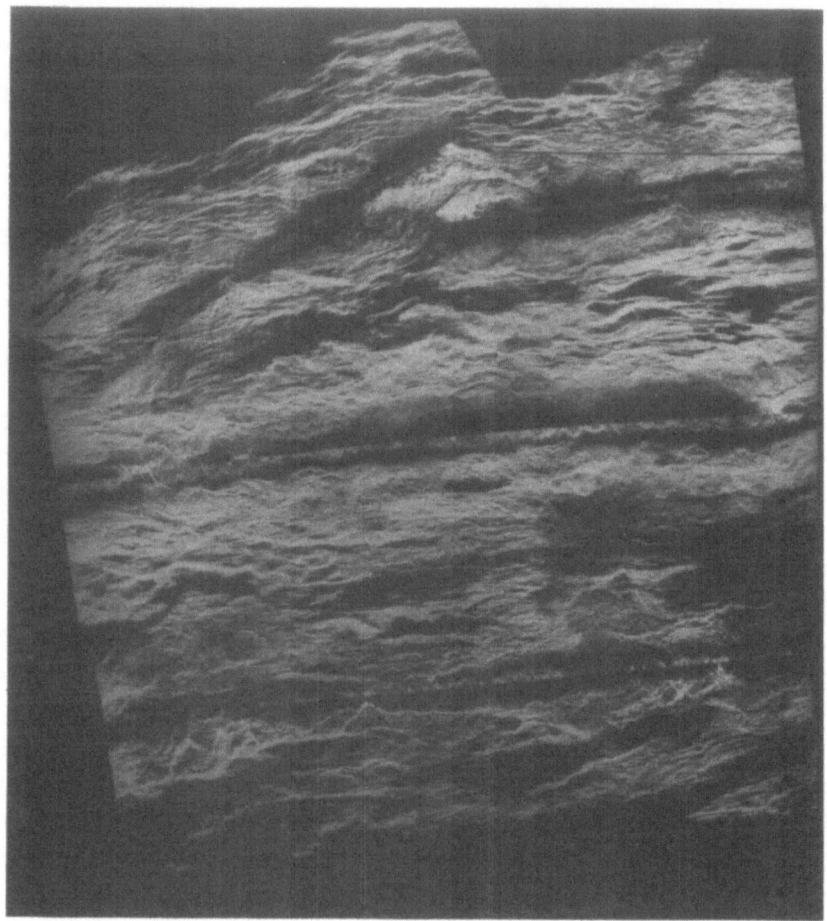

Fig. 11. Digitally mosaicked GLORIA images. Mosaic is approximately 100 km square, straddling the axis of the slow-spreading Southwest Indian Ridge near 27.5° S, 66° E.

14 m which is unlikely to affect the geological interpretation, and is in fact close to the real resolution of the Sea Beam system (Renard and Allenou, 1979). A colour scale is then attached to these 256 values, each colour being represented by predefined ratios of red to green to blue. We use a table of colour values that produce a 'rainbow' colour succession, from red for shallowest depths to violet for deepest.

The GLORIA and Sea Beam information are combined by noting that the colour and brightness of any image pixel can be defined in terms of the values of three independent primary colours: red (r), green (g), and blue (b). The hue of the pixel is determined by the three ratios $r{:}g{:}b$, and its brightness by their absolute values. We use the hue to represent the depth (coded from the Sea Beam pixel value), and the intensity to represent the GLORIA value. In principle one would expect the brightness to be a simple sum $r + g + b$; however, because both the

TABLE II

Colour table used in producing Figs 11a and b

Level	Red	Green	Blue	Approximate colour
1	255	0	255	Deep purple
42	204	0	10	Deep red
104	255	255	0	Yellow
126	0	204	0	Green
168	0	255	255	Cyan
210	0	0	204	Deep blue
255	204	0	204	Magenta

Intermediate levels were produced by linear interpolation between these values.

sensitivity of the eye to different colours, and the strengths of different CRT phosphors, photographic dyes, and plotting inks, are very variable, different values of the sum $r + g + b$ must be used to give the same perceived brightness for different colours. To

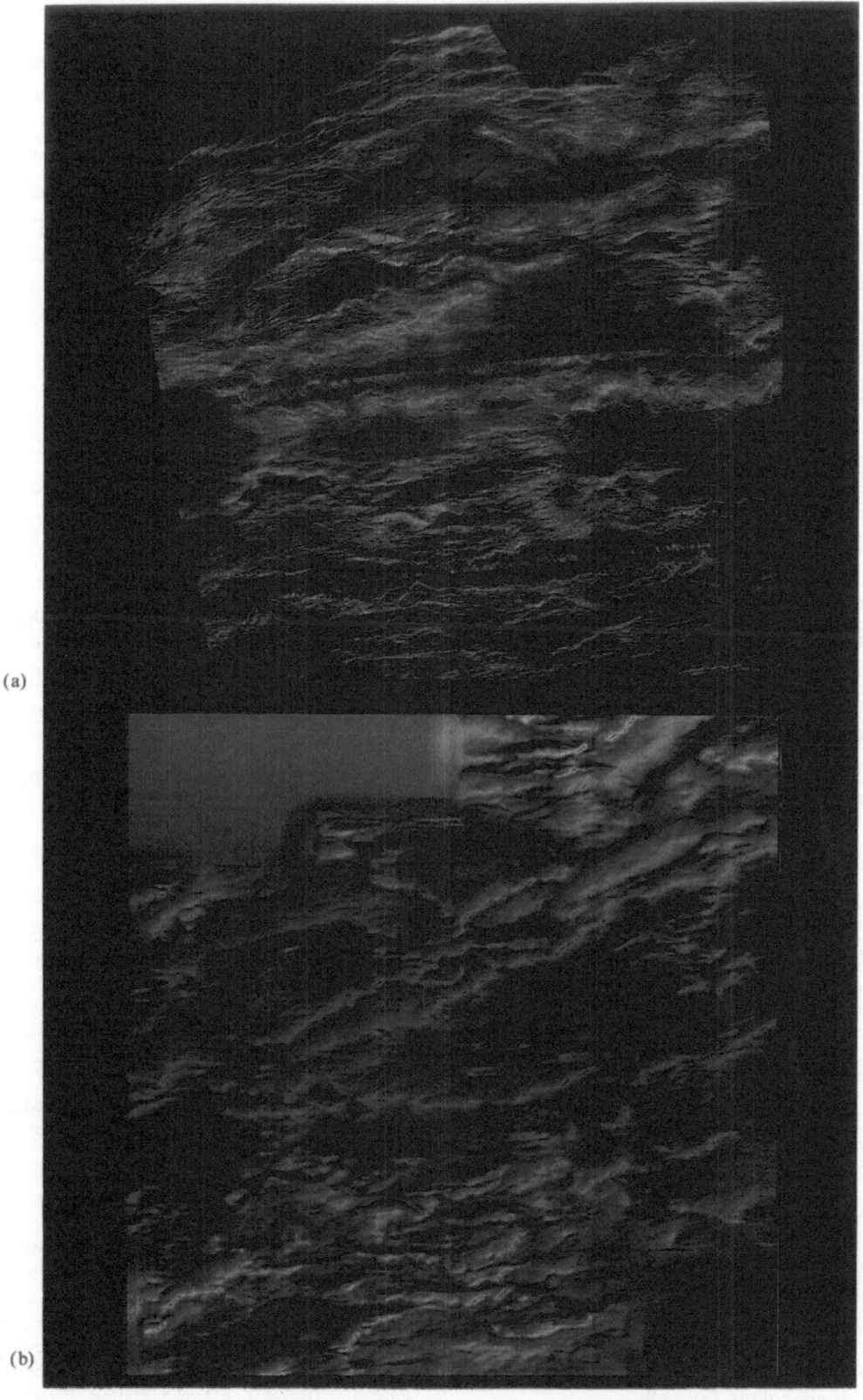

(a)

(b)

Fig. 12. (a) (above) coloured image of same area as Fig. 11, made by combining sidescan (GLORIA) and depth (Sea Beam) information; the hue is a function of depth, while the intensity is a function of sonar backscatter. Blue (deep) areas near the centre of the image are *en échelon* basins marking the spreading axis. Note that the addition of depth information aids greatly in "reading" the image. (b) (below) shaded relief image of the area shown in (a), for comparison, made from bathymetric (Sea Beam) information alone. The image intensity represents the amount of light that would be reflected from the given topography for an illumination source 20° above the horizon to the south of the area, assuming reflection according to Lambert's Law; image colour is related to depth. Note that (a) shows finer detail than (b) (and also some textural information, though that is less important in this region where overall sediment cover is low).

some extent the choice of levels must therefore be made subjectively. Table II gives the $r\,g\,b$ table used for the examples shown in this chapter.

As an example, assume a Sea Beam depth of 3000 m and a GLORIA intensity of 47. The position in the colour table is $(3000-2000)*256/(5500-2000) = 73$. Colour value 73 is predefined as the ratio $r{:}g{:}b = 230{:}128{:}5$, a bright (at maximum intensity) yellowy-red (see Table II). Therefore the required output values are:

$$r = 230 \times 47/256 = 42;$$
$$g = 128 \times 47/256 = 24;$$
$$b = 5 \times 47/256 = 1.$$

Total output is therefore 42:24:1, a dark yellowy-red.

The resultant Sea Beam/GLORIA image is shown in Fig. 12a. The positions of structural highs and lows in relation to sonar information can now be seen at a glance. The combined image also helps explain the presence of certain bright or dull spots, as being due either to the angle of insonification or to real lithological variation. It gives a somewhat similar appearance to a shaded relief map constructed just from Sea Beam data (Figure 12b), but the latter can give no information on lithology or surface roughness, and also lacks the resolution to reveal some of the very fine detail seen in the GLORIA data. This is clearly a powerful technique for combining different data sets, with all geographical cross-referencing having been incorporated into the image.

Although we have not yet tried them, many other possibilities for combining data by this method exist. One could reverse the combination, and relate image hue to sidescan intensity (or, better, to the derived backscattering coefficient), and image intensity to topography in a shaded relief representation. The method would be equally suitable for combining sidescan or topography with, for example, gravity and magnetic fields, or sediment thickness, to great advantage. It would even be possible to combine more than two types of data, by using different intensities of a single colour for each.

Conclusions

A suite of programs is now available for applying geometric and radiometric corrections to GLORIA and other sidescan images, most of which can be used in near real time on board ship. Other programs provide enhanced display of images, either alone or combined with other data sets. Finally, a start has been made in recovering true values of the acoustic backscattering coefficient. These developments are the latest in a series of continuous improvements to the system over a period of twenty years. Together with efficient and reliable digital data acquisition, they now enable GLORIA to be used as a truly quantitative research and survey tool.

Acknowledgments

We gratefully acknowledge the help and advice of all those who have been involved in developing and operating the GLORIA system over the years. We are especially thankful to Pat Chavez and his colleagues in USGS, not only for making the MIPS software available to us, but for many hours spent with us explaining it and image processing in general. We also thank Angela Morrison, John Baker, and other staff at the NERC Thematic Information Servies for helping us in our early struggles with the I^2S system.

References

Chavez, P. S., 1986, Processing Techniques for Digital Sonar Images from GLORIA, *Photogrammetric Engineering and Remote Sensing* **52**, 1133–1145.

EEZ-Scan 84 Scientific Staff, 1988, Physiography of the Western United States Exclusive Economic Zone, *Geology* **16**, 131–134.

Laughton, A. S., 1981, The First Decade of Gloria, *J. Geophys. Res.* **86**, 11511–11534.

McKinney, C. M. and Anderson, C. D., 1964, Measurements of Backscattering of Sound from the Ocean Bottom, *J. Acoust. Soc. America* **36**, 158–163.

Reed, T. B., 1987, *Digital Image-Processing and Analysis Techniques for Sea MARC II Side-Scan Sonar*, Ph.D. thesis, University of Hawaii.

Renard, V. and Allenou, J. P., 1979, Seabeam, Multi-beam Echo-Sounding in "Jean Charcot", *Int. Hydr. Rev.* **56**, 35–67.

Rusby, J. S. M., 1970, A Long-Range Side-Scan Sonar for Use in the Deep Sea (GLORIA Project). *Int. Hydrogr. Rev.* **47**, 25–39.

Rusby, J. S. M. and Somers, M. L., 1977, The Development of the "Gloria" Sonar System from 1970 to 1975, in M. Angel (ed.), *A Voyage of Discovery*, 611–625, Oxford, Pergamon (Suppl. to Deep-Sea Research).

Searle, R. C. and Hey, R. N., 1983, GLORIA Observations of the Propagating Rift at 95.5° W on the Cocos-Nazca Spreading Center. *J. Geophys. Res.* **88**, 6433–6447.

Searle, R. C. and Hunter, P. M., 1986, The Use of GLORIA Long-Range Sidescan Sonar for Deep-Ocean Mapping, in M. Blackmore (ed.), *Autocarto London, 2, Digital Mapping and Spatial Information Systems*, 339–388.

Searle, R. C. and Kidd, R. B., 1984, GLORIA Digital Image Processing and Interpretation of the Saharan Sediment Slide, Eastern North Atlantic, *EOS, Trans. Amer. Geophys. Union* **65**, 1082–1083 (Abstract).

Somers, M. L., 1970, Signal Processing in Project GLORIA, A Long Range Side-Scan Sonar, in D. G. Tucker (ed.), *Electronic Engineering in Ocean Technology*, Proc. Inst. Electronic and Radio Engineers Conference, Swansea, 1970, Institution of Electronic and Radio Engineers, London, 109–120.

Somers, M. L., 1973, Some Recent Results with a Long-Range Side-Scan Sonar, in *Signal Processing* (Proceedings of the NATO Advanced Study Institute, Loughborough, 1972), Academic Press, London, 757–767.

Somers, M. L., Carson, R. M., Revie, J. A., Edge, R. H., Barrow, B. J., and Andrews, A. G., 1978, GLORIA II—An Improved Long-Range Side-Scan Sonar, in Proc. IEEE/IERE Subconference on Offshore Instrumentation, *Oceanology International '78*, Technical Session J, pp. 16–24, BPS Publications, London.

Urick, R. J. 1975, *Principles of Underwater Sound for Engineers*, 2nd edition. McGraw-Hill, New York.

High-Resolution Seismic Reflection Surveying of Shallow Marine and Estuarine Environments

J. M. REYNOLDS

Department of Geological Sciences, Polytechnic South West, Drake Circus, Plymouth, PL4 8AA, UK

(Received 27 April, 1989; accepted 1 September, 1989)

Key words: Seismic reflection profiling, geophysical interpretation methods.

Abstract. High-frequency seismic reflection profiling is a well-established and often used technique in marine investigations. Traditionally seismic data are viewed as two-dimensional time sections. Given closely spaced profile lines, it is possible to produce posted two-way travel time maps of sub-surface reflectors which, when plotted as isometric displays, clearly show the three-dimensional spatial morphology of the sub-surface topography. With borehole control, such information can be used to provide a series of images which indicate temporal as well as spatial relationships of sub-surface reflectors. With the high-resolution afforded by high-frequency methods, detailed information on palaeo-environments can be reconstructed. Using the geophysical database as a basic framework, other aspects of the same environment can be examined in considerable detail. To demonstrate the effectiveness of these procedures, examples will be given from Plymouth Sound where a series of nested buried rock valleys has been mapped in detail from Sparker and Boomer surveys which have been interpreted in the light of newly-acquired borehole information. Isometric plots of the various sub-surface interfaces show how the channels have developed as sea level has risen over the last ca. 10,000 years. Different sedimentological facies can be resolved within the channel system thus providing information about the processes involved in their formation. Methods of improving the resolution and subsequent geological interpretation of high-resolution seismic reflection surveys are being developed for shallow marine and estuarine environments such as those found in Plymouth Sound.

Introduction

Although high-frequency seismic reflection profiling is a well-established and often-used technique in marine site-investigations, the guidelines (British Standard 5930; 1981) for its application are notoriously inadequate. The question "What constitutes an *adequate* interpretation of high-resolution seismic data?" has still to be addressed. Nevertheless, the method is commonly employed in a variety of appli-

cations (Table I) by way of three types of seismic source systems, namely, Pingers, Boomers and Sparkers. More details of the types of sources are given by Sieck and Self (1977), Lugg (1979) and McQuillin *et al.* (1984), and will be discussed in more detail later. The main functions of seismic profiling in these applications are two-fold: to define basic engineering and geological parameters, and to discover any potentially hazardous sub-surface geological conditions. Examples of the former are rippability parameters, overburden thickness, and the recognition of simple rock types. Examples of the latter include seabed pockmarks (which may indicate degassing of sediments), buried rock valleys (which may cause problems with foundation stability and drilling), gasified sediments (low-pressure gas) which would be problematical if large structures were constructed over them, and gas accumulation (high-pressure gas) which can cause blowouts during drilling (as for example has happened in the Gulf of Mexico). This chapter will concentrate more specifically on applications to the near-shore and estuarine environments with water depths in the general range 0–25 m.

One of the reasons why high-resolution seismic profiling is such an attractive option in commercial offshore site investigation is that it provides a cheap and effective way of obtaining images of the subsurface. A typical single-channel analogue acquisition system produces a graphical output for which subsequent data processing is rarely possible, but which can yield much valuable geotechnical information. The seismic sections produced are two dimensional images of the geological environment at the time of the survey. For many applications the level

TABLE I

Principal applications of commercial offshore site-investigation using high-resolution seismic methods. (Modified from Carter *et al.*, 1986)

Near-shore marine/estuarine environments
   Bridges, tunnels, viaducts
   Harbours, jetties, quay walls, marinas
   Pipelines and tunnels for sewage outfalls
   Dredging for access channels to ports/harbours
Marine
   Hydrocarbon pipelines
   Hydrocarbon production platforms/well heads
   Siting of drilling rigs

of interpretation which this affords may well be acceptable. For others it is not, and one of the objectives of this chapter is to demonstrate that single-channel analogue seismic records can provide a four-dimensional spatial and temporal model of the sub-surface geological environment. The information so obtained can be of immense value in such areas as environmental impact assessment investigations and hydrogeological studies, as well as in establishing the nature of sedimentary sequences, all of which can aid engineering design. It is important to recognize that there are sophisticated digital multichannel high-resolution seismic systems available, the data from which can be enhanced using commercial processing packages, and which, consequently, are very expensive to use. The more usual site-investigation practice in near-shore and estuarine environments is to employ the low-cost single-channel analogue graphical systems. It would be an advantage to be able to improve the level of geotechnical interpretation of data acquired by these analogue systems without incurring a prohibitive increase in costs. The purpose of this chapter is to demonstrate that judicious survey design, coupled with a limited amount of computer power, can enhance greatly the information obtainable from these basic surveys and provide a very useful foundation for further investigations.

The Department of Geological Sciences and the Institute of Marine Studies at Polytechnic South West have undertaken high-resolution seismic reflection surveys within Plymouth Sound and the River Tamar since 1982 (McCallum and Reynolds, 1987; Reynolds, 1987). Following financial support from the National Advisory Body, a commercial jack-up

rig drilled two boreholes within Plymouth Sound in February 1988. Examples from the work in Plymouth Sound will be used in this chapter to illustrate particular aspects of seismic interpretation.

Survey Constraints

POSITION FIXING

As with any survey, navigation is of vital importance. The accuracy of the position fixing methods employed will constrain the spatial interpretation of the seismic data. If the ground conditions vary markedly over short distances (e.g. 30 m), it would prove an almost impossible task to locate a borehole over a specific spot if the positions of the survey stations were known to no better than, say, 20 m. Within Plymouth Sound, for example, the depth to bedrock beneath the mudline within the estuary can vary from about 3 m to 20 m in a horizontal distance of only 50 m. For positioning the seismic lines and the two boreholes of this survey, Trisponder systems were used. These resulted in a standard deviation on position of between 0.5 and 1 m. In the future, it is likely that positional accuracies to within ± 1 m will become more commonplace once the Global Positioning System (GPS) becomes fully established (Bullock, 1988). This should help to solve the problems associated with uncertainties in position, permit closer line spacings (e.g. 5 to 10 m separations) and allow greater opportunities for spatial interline interpretation of seismic sections.

SOURCE PARAMETERS

One of the factors which controls the resolution of a seismic survey is the shape of the source pulse. Ideally this should be a spike (Dirac) pulse where all the energy is applied at a single instant of time. In reality, best resolution occurs with those sources which produce a wave shape as close to the ideal as possible, typically a minimum-phase wavelet (Waters, 1978, p. 164). A seismogram, denoted by s, comprises the convolution of the source wave w with e, a time series which contains the acoustic impedance function of the media through which the signal is propagating such that:

$$s = w*e(+n),$$

where * denotes convolution and n, additive noise (Hatton *et al.*, 1986). Consequently, the resolution of

TABLE II

Theoretical resolution and depth penetration of three common high-frequency seismic sources

Source	Frequency bandwidth	Resolution	Depth of penetration
Pingers	3.5–7 kHz	0.1–1.0 m	≤ tens of metres
Boomers	400 Hz–5 kHz	≈ 1 m	tens to 100+ m
Sparkers	200 Hz–1.5 kHz	2–3 m	≥ 1000 m

a seismic section is dependent upon the quality of the source wavelet w.

Parameters of the source wavelet which particularly affect the quality of the seismic sections are the pulse shape and the pulse length.

PULSE SHAPE

The higher the source frequency, the shorter the wavelength of the signal and hence (in theory) the better the resolution. However, high-frequency signals are attenuated more rapidly than those at lower frequencies, so that the higher the source frequency, the shallower is the depth penetration (Table II). However, such specification summaries (e.g. Conway et al., 1986) are misleading. With Sparkers, for example, the pulse shape depends critically on the output power of the system and the number and configuration of electrodes in the sparker array. Examples of this source variability are given by Lugg (1979) and are illustrated in Fig. 1. Better pulse shapes are achieved by increasing the number of electrode tips. Within Plymouth Sound, a 3-tip Sparker array at 500 J produced significantly broader pulses with a corresponding decrease in resolution than did an 18-tip array at the same output energy. In addition, the source depth (below the sea surface) also affects the pulse shape by virtue of the interference of the surface ghost and bubble pulse. The effect of these parameters is to cause the source wavelet to become a complex non-ideal shape and, furthermore, to differ from the manufacturer's specification for the

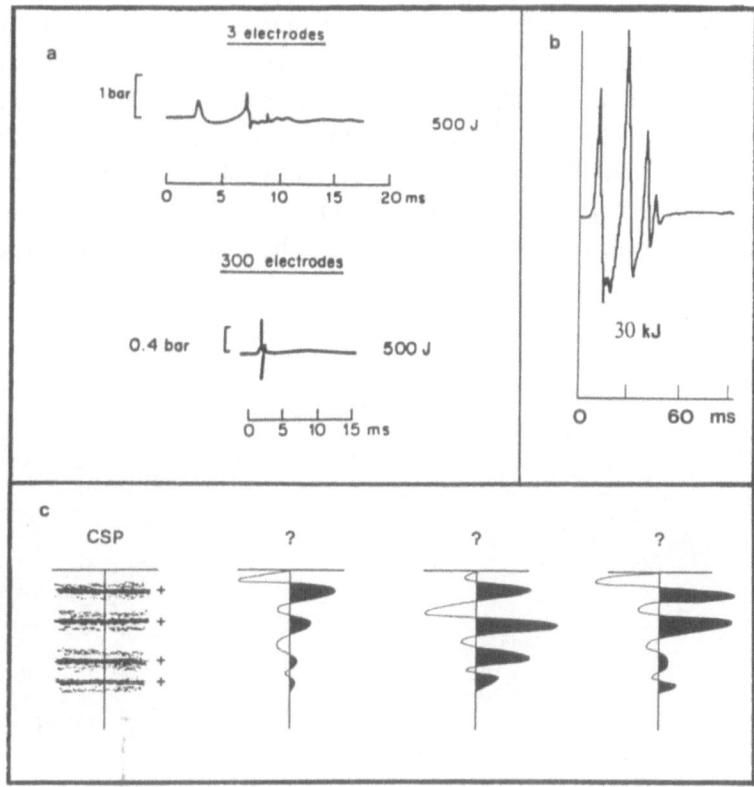

Fig. 1. Sparker source pulse shapes with (a) 3 and 300 electrode tips with 500 J output (after Lugg, 1979), and (b) a far-field signature for a system with 30 kJ output (after Hatton et al., 1986). (c) Analogue signal form from Plymouth Sound (left) with pulse polarity but with no amplitude information, and right, a variety of waveforms with the same pulse polarity and phase but different amplitudes and which could represent the observed signal.

device used, because of the dependence of performance on the local ambient conditions of operation.

PULSE LENGTH

The ability to resolve between reflectors in an interpretation depends upon the pulse length (Meckel and Nath, 1979), the amount of reverberation which occurs and on the distribution and nature of the sub-surface media. If a pulse length of, say, 5 ms is used and the two-way travel time difference between two events is greater than the pulse period, then it should be possible to resolve each reflection. However, if reverberation of the pulse occurs and the tail of the first reverberation obscures the onset of the second event and interferes with it, then the second reflection may remain unresolved and effective deconvolution of the acoustic impedance function of the sub-surface media is unsuccessful. It is therefore to be recommended that the source pulse shape and duration should be recorded at the time of each survey. The observed pulse shape can then be used, if necessary, in the generation of any synthetic seismograms to aid the interpretation. Another factor is the nature of the media through which the acoustic pulses travel. Penetration through and the seismic character of specific types of material also affect the choice of source. Some indication of this is given by Carter *et al.* (1986). Within Plymouth Sound, for example, and EG & G Boomer produced more smeared seismic sections and poorer resolution than

a multi-tip Sparker at the same energy output (0.5 kJ). The quality of the Sparker data was sufficiently good that the main geological interfaces predicted from the seismic interpretation were all confirmed to within 0.5 m by subsequent drilling. It was also found that the Sparker results degraded slightly throughout the survey as some of the electrode tips burned back to the insulation and eventually ceased to work, thereby reducing the number of active tips and the effective output both of which served to degrade the pulse shape. It is important to check on the state of electrode tips during each day's survey and to monitor signal quality.

Detailed Seismic Interpretation

Traditionally, analogue seismic data are viewed as two-dimensional time sections which can be interpreted simplistically in terms of line drawings of the principal reflections. This level of interpretation is commonly used in non-critical site investigations where only the depth to bedrock is of importance and detailed interpretation is not required. It is possible to go a stage further and interpret the sections in terms of seismic stratigraphic sequences. Despite the apparent relative coarseness of the resolution of Sparker data, it has been possible to define detail within a sequence of sediment progradational surfaces with a time resolution of less than 1 ms (Fig. 2). The sigmoidal form of the reflections is very

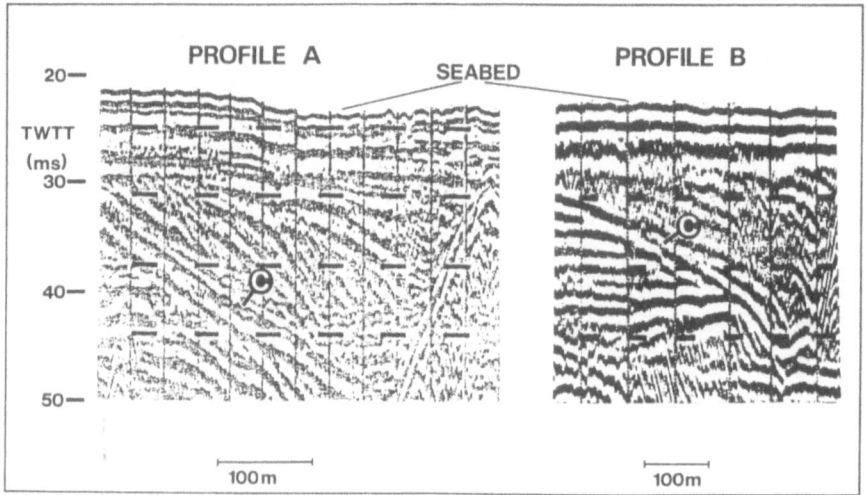

Fig. 2. Two extracts from sparker records in the range 20–50 ms Two-Way Travel Time (TWTT) of the same ground within Plymouth Sound showing sigmoidal reflections associated with a known sand body (after Eddies and Reynolds, 1988). The two records were acquired with different filter frequency bandwidths.

similar to that of reflection events described by Mitchum *et al.* (1977). Where low-frequency multichannel data have been recorded digitally the sophistication of interpretation is well developed (e.g. Todd and Mitchum, 1977; Brown *et al.*, 1984), but with analogue records the options for further interpretation are much more restricted. Nevertheless, given a sufficient number of closely spaced seismic profile lines, such as the one illustrated in Fig. 3, it is possible to produce posted Two-Way Travel Time (TWTT) maps (e.g. Fig. 4) of sub-surface reflectors which, when plotted as isometric displays (Fig. 5), clearly show the three-dimensional spatial morphology of the sub-surface topography. Figure 5 illus-

trates stacked isometric projections of Two-Way Travel Times, which have been normalized to a time datum of 60 ms below Ordnance Datum (Newlyn) for graphical display. Four surfaces have been selected: seabed, reflector 2, reflector 1 and bedrock. Reflector 1 correlates with the shelly gravel horizon (Fig. 3) which separates organic muds from overlying sands. Reflector 2 is a prominent event associated with the sigmoidal reflections seen in Figs 2 and 3 and has been interpreted as being a point bar foreset (Eddies and Reynolds, 1988). Two Way Travel Time maps and isometric projections were produced by digitizing line interpretations of the seismic sections and processing the data using the SURFER graphi-

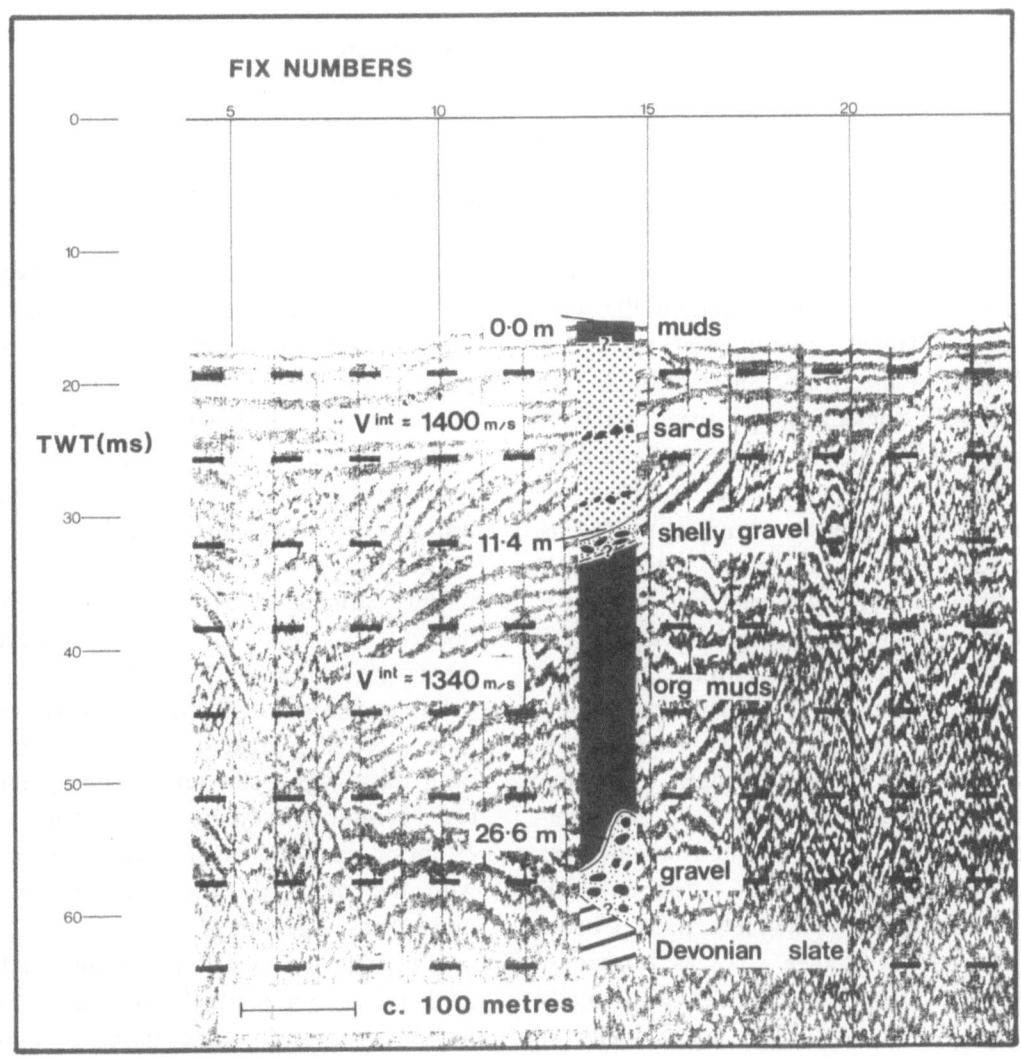

Fig. 3. Analogue Sparker record over a buried rock valley within Plymouth Sound with the main geological sequences as found by drilling. Two preliminary average interval velocities have been determined for the main sequences corresponding to the depth ranges 0–11.4 m and 11.4–26.6 m using the observed Two-Way Travel (TWT) Times and the measured depths to the corresponding interfaces from the borehole lithological log.

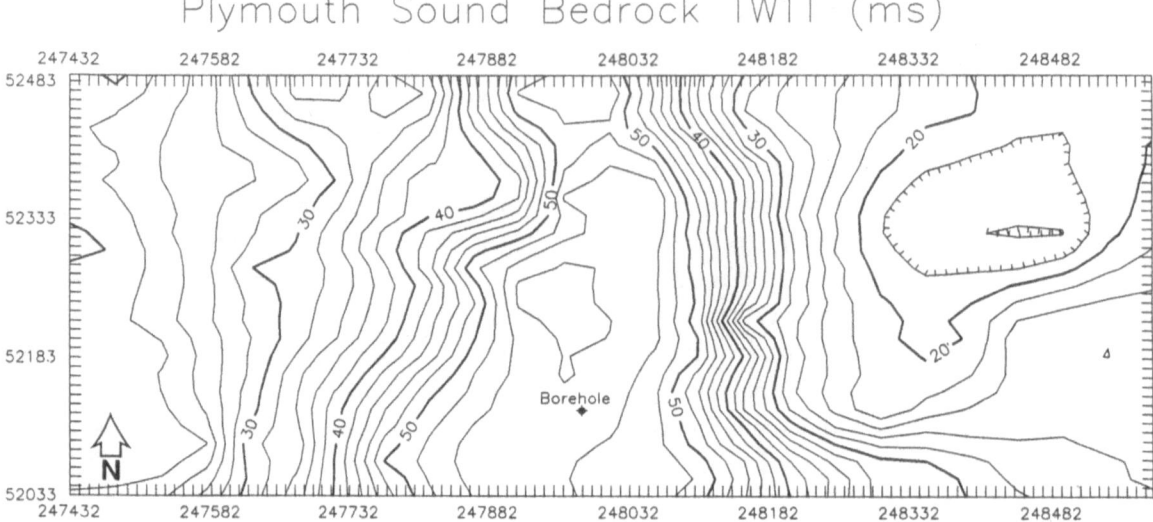

Fig. 4. An example of a Two-Way Travel Time map to bedrock from a part of Plymouth Sound. Values along the X and Y axes are Universal Transverse Mercator (UTM) co-ordinates. The location of the borehole (details in Fig. 3) is also shown.

cal package (Golden Software, Inc., Colorado, USA) on an IBM PC-XT compatible microcomputer. Hardcopy output was produced via a 24-pin printer. The stacked isometric projections illustrate the upward and lateral migration of the main drainage channel of the River Tamar through Plymouth Sound as a consequence of rising sea level and subsequent sedimentation. With borehole control, such information can be used to provide a series of images which indicate temporal as well as spatial relationships of sub-surface reflectors. A borehole with good recovery of material sited on a well-interpreted seismic line provides a means for calibrating the seismic sections. More than that, interdisciplinary investigations of the retrieved material can provide information about when certain events may have occurred. For example, within the cores from Plymouth Sound, extremely well-preserved micro-fossils have been found which have been used to date major horizons. These, in turn, have been correlated with major reflections. Consequently, some reflections then serve as isochrons. For example, one horizon (top of the organic muds shown in Fig. 3) has been dated provisionally at 8300 BP (Eddles and Hart, in press) on the basis of microfossils; similar mid-Flandrian ages have been obtained from samples of the same material from preliminary palynological investigations (J. Scourse, pers. comm.). This can give some indication as to

process rates within the sedimentological regime as well as to how the dynamics of a particular environment relate to larger scale climatic events such as sea level changes (Eddies and Reynolds, 1988). Although only provisional dates are available at present, the core material is currently being sub-sampled, not only for further detailed micro- and macro-palaeontological, palaeo-botanical and palynological studies, but also for heavy metal concentrations, clay mineralogy, and organic geochemistry. With the high-resolution afforded by single-channel seismic profiling methods (Table II), it should be possible to develop more detailed palaeo-environmental reconstructions.

Survey techniques in deeper water and on land have received a great deal of attention in terms of both data acquisition and in interpretation. However, seismic surveys in very shallow marine and estuarine environments, especially those associated with the intertidal zone, are assuming increasing importance. Such an environment is a very difficult one in which to work logistically and which fits neither land nor marine surveying techniques comfortably. However, an example of a three-dimensional seismic survey over tidal mudflats in The Netherlands has been described recently by Corsmit et al. (1988). The success of the survey was dependent upon (a) the acquisition of multifold coverage and (b) data processing on a minicomputer.

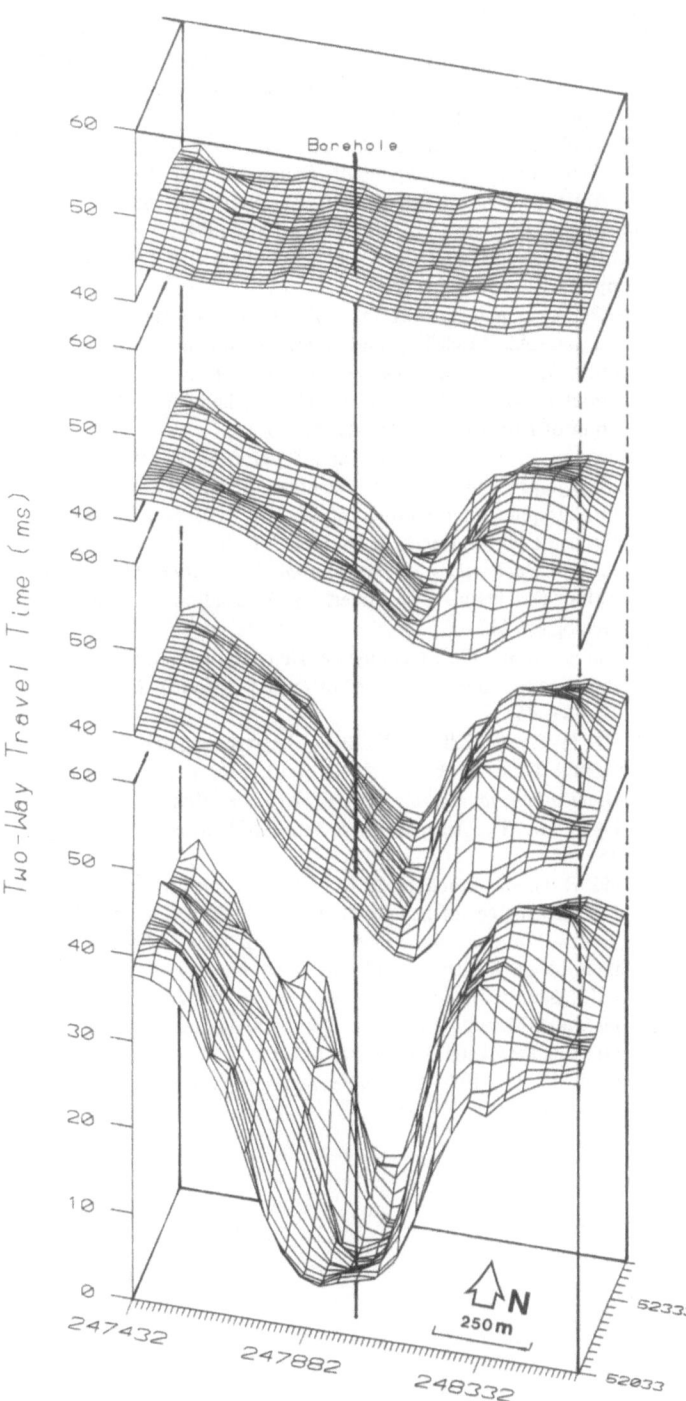

Fig. 5. Examples for part of Plymouth Sound of stacked isometric projections of normalised Two-Way Travel Times to seabed, reflector 2, reflector 1 and bedrock respectively. Values along the X and Y axes are Universal Transverse Mercator (UTM) co-ordinates. The location of the borehole (details in Fig. 3) is also shown.

Methods for improving the resolution and subsequent geological interpretation of shallow seismic reflection surveys are being developed at the Department of Geological Sciences at Polytechnic South West, Plymouth, for shallow marine and estuarine environments such as those found in Plymouth Sound. Such developments will be of benefit to hydrographers, geologists and civil engineers as well as to mariners and environmental conservationists.

Conclusions

Factors which are known to constrain the quality of data acquired during an analogue seismic profiling survey include: navigational accuracy, the shape of the seismic source wavelet, and the nature and geometry of the sub-surface reflectors.

Analogue seismic records are usually interpreted two-dimensionally and with only the main reflections being picked out on line-drawing representations. However, the use of seismic stratigraphic sequence interpretation allows inferences to be drawn concerning the dynamics of the depositional regime and the processes involved in producing the resultant sub-surface geology. This in itself does not go far enough. The use of a sufficient number of closely-spaced seismic survey lines permit the definition of isometric surfaces of the principal reflections and thus the production of a three-dimensional model of the spatial distribution of particular seismic events. In addition, by modelling events from the rockhead through to the sea/river floor it is possible to obtain an impression of both the spatial and temporal relationship between features. Allied with borehole control which can give probable dates for specific events, a dynamic model of the sub-surface regime can be developed. This approach can be of great value in aiding the geotechnical, hydrogeological and geological interpretation of analogue seismic data, and it offers the potential of aiding civil engineering design in the evaluation of possible post-development environmental problems.

Acknowledgements

I am grateful to Capt. K. McCallum, Mr. C. Sawyer, and Mr. T. Parrott, all of the Institute of Marine Studies at Polytechnic South West, Plymouth, for their collaboration and assistance, without which the field work would not have been possible. The Plymouth Sound Project has received valuable financial support from the National Advisory Body. Thanks are also due to Mr. R. D. Eddies for fruitful discussions concerning the seismic work and for his help in drafting some of the diagrams, and to Dr E. Hailwood for his many editorial improvements.

(Plymouth Sound Research Project Contribution No. 8).

References

BS 5930, 1981, *Code of Practice for Site Investigations* (Formerly CP 2001), British Standards Institution.

Brown, A. R., Wright, R. M., Burkart, K. D., Abriel, W. L., and McBeath, R. G., 1984, *Tuning Effects, Lithological Effects and Depositional Effects in the Seismic Response of Gas Reservoirs.* Presented at the 46th Annual Meeting of the European Association of Exploration Geophysicists, London, June 21, 1984.

Bullock, S. J., 1988, Future and Present Trends of Navigation and Positioning Techniques in Exploration Geophysics, *Geophysical Journal* **92**(3), 521.

Carter, P. G., Pirie, R. M., and Sneddon, M., 1986, Marine Site Investigations and BS 5930, in Hawkins, A. B. (ed.), *Site Investigation Practice: Assessing BS 5930*, Geological Society Engineering Geology Special Publication No. 2, 163–66.

Conway, B. W., McCann, D. M., Sarginson, M., and Floyde, R. A., 1984, A Geophysical Survey of the Crouch/Roach River System in South Essex with Special Reference to Buried Channels, *Quarterly Journal of Engineering Geology* **17**(3), 269–82.

Corsmit, J., Verteeg, W. H., Brouwer, J. H., and Helbig, K., 1988, High-Resolution 3D Reflections on a Tidal Flat: Acquisition, Processing and Interpretation. *First Break* **6**(1), 9–23.

Eddies, R. D. and Reynolds, J. M., 1988, Seismic Characteristics of Buried Rock Valleys in Plymouth Sound and the River Tamar, *Proceedings of the Ussher Society* **7**, 36–40.

Eddles, A. P. and Hart, M. B., 1989, Late Quaternary Foraminiferida from Plymouth Sound: Preliminary Investigation, *Proceedings of the Ussher Society* **8** (in press).

Hatton, L., Worthington, M. H., and Makin, J., 1986, *Seismic Data Processing*, Oxford, Blackwell Scientific Publications, 177 pp.

Lugg, R., 1979, Marine Seismic Sources, in Fitch, A. A. (ed.), *Developments in Geophysical Exploration Methods — 1*, London, Applied Science Publishers Ltd, 143–203.

MacCallum, K. D. and Reynolds, J. M., 1987, High-resolution Seismic Profiling in Plymouth Sound and the River Tamar (Abstract), *Proceedings of the Ussher Society* **6**, 562.

McQuillin, R., Bacon, M., and Barclay, W., 1984, *An Introduction to Seismic Interpretation*, London, Graham and Trotman, 287 pp.

Meckel, L. D. and Nath, A. D., 1977, Geologic Considerations for Stratigraphic Modelling and Interpretation, in Payton, C. E. (ed.), *Seismic Stratigraphy — Applications to Hydrocarbon Exploration*, Memoir 26, Tulsa, Oklahoma, The American Association of Petroleum Geologists, 417–38.

Mitchum, R. M., Vail, P. R., and Sangree, J. B., 1977, Seismic Stratigraphy and Global Changes of Sea Level, Part 6: Stratigraphic Interpretation of Seismic Reflection Patterns in Depositional Sequences, in Payton, C. E. (ed.), *Seismic Stratigraphy — Applications to Hydrocarbon Exploration*, Memoir 26, Tulsa, Oklahoma, The American Association of Petroleum Geologists, 117–33.

Reynolds, J. M., 1987, Geophysical Detection of Buried Channels in Plymouth Sound, Devon (Abstract), *Geophys. J. Roy. Astron. Soc.* **89**, 457.

Sieck, H. C. and Self, G. W., 1977, Analysis of High Resolution Seismic Data, in Payton, C. E. (ed.), *Seismic Stratigraphy — Applications to Hydrocarbon Exploration*, Memoir 26, Tulsa, Oklahoma, The American Association of Petroleum Geologists, 353–85.

Todd, R. G. and Mitchum, R. M., 1977, Seismic Stratigraphy and Global Changes of Sea Level, Part 8: Identification of Upper Triassic, Jurassic, and Lower Cretaceous Seismic Sequences in the Gulf of Mexico and Offshore West Africa, in Payton, C. E. (ed.), *Seismic Stratigraphy — Applications to Hydrocarbon Exploration*, Memoir 26, Tulsa, Oklahoma, The American Association of Petroleum Geologists, 145–63.

Waters, K. H., 1978, *Reflection Seismology*, New York, John Wiley and Sons, 377 pp.

A Fixed Receiver for Recording Multichannel Wide-Angle Seismic Data on the Seabed

C. M. R. ROBERTS

Department of Geological Sciences, University of Durham, UK

and

M. C. SINHA

Bullard Laboratories, Department of Earth Sciences, University of Cambridge, UK

(Received 27 April, 1989; accepted 1 September, 1989)

Key words: seismic refraction, lithosphere, continental shelf.

Abstract. Previous experiments to record seismic data at wide angle on the continental shelf have generally been unsuccessful in determining velocity structure in the lower crust; either the lines were too short or shot-receiver density too sparse to identify lower crustal arrivals. In contrast, deep normal incidence profiles show good structural resolution in the crust and uppermost mantle. A sea-bottom multichannel instrument has been developed to record datasets containing closely spaced traces, in order to improve the resolution of reversed wide-angle experiments on the continental shelf.

The Pull-up Multichannel Array (PUMA) is a 1200 m, 12-channel hydrophone array for remotely recording seismic data on the seabed. It consists of 12 short hydrophone sections linked by 100 m-long passive sections. A pressure case is attached to the array at one end, in which recording electronics, cassette tape recorders and a battery power supply are housed. The PUMA is designed for deployment in water depths less than 200 m from a research ship and is moored to buoys for recovery.

The instrument, which was successfully used in an experiment west of Lewis, Outer Hebrides, UK (Powell and Sinha, 1987) was specifically designed to provide a reliable determination of the velocity structure of the crust and uppermost mantle over part of the BIRPS WINCH deep normal incidence profile. Because the traces are closely spaced it is easy to correlate phases across the record section and to monitor changes in amplitude. A velocity structure for the continental crust and uppermost mantle has been devised from these data, using amplitude modelling.

Introduction

The purpose of this chapter is to show that it is possible to achieve significant improvements in the resolution of conventional wide-angle seismic experiments on the continental shelf by recording multi-channel data from a fixed sea-bottom receiver. Compared with earlier techniques, this method provides a closer determination of both the velocity structure and the composition of the continental lithosphere, in particular the lower continental crust and Moho.

Currently used seismic techniques at both normal incidence and wide-angle are first reviewed. Existing seismic experimental methods are shown to have certain limitations when used on the continental shelf. A new instrument is then described which has been used to overcome these disadvantages by recording more closely-spaced data at wide angle on the seabed.

Review of Existing Normal Incidence and Wide-Angle Seismic Techniques

The resolution of structures within the continental lithosphere has greatly increased with the development of deep normal incidence seismic profiling (Meissner *et al.*, 1983). Phinney (1978) estimates that the current resolution of deep reflection profiling is approximately 150 m vertically and 500 m horizontally, probably an order of magnitude smaller than the uncertainties in crustal structure determined from seismic experiments at wide angle.

Deep reflection data are good for imaging deep structure, but give no quantitative insight into the seismic velocities of the deep crust because of the small amount of moveout across the receiver array. Wide-angle experiments, although suffering from poorer structural resolution, do provide detailed velocity–depth information which can be related to experiments on material taken from exposures of crystalline basement for determining crustal composition.

Information derived from seismic studies at wide-angle and normal incidence is complementary, and so there has been an increasing number of combined normal incidence and wide-angle experiments (e.g. Mooney and Brocher, 1986). Many of these show good correlations between the two datasets (Mueller, 1977; Barton et al., 1984). When both normal incidence and wide-angle techniques are used at the same locality, the depths of reflectors observed on normal incidence profiles can be calculated using velocity–depth structures determined at wide angle. Only then can the position of the Moho or other structures of interest be identified on normal incidence records. Accurate determinations of velocity are also required for migrating deep reflection data.

During the last fifteen years major advances have been made in wide-angle seismic deep sounding to improve the resolution of this technique. Interpretation of wide-angle datasets has evolved from one-dimensional models representing two or three layers, to complex, laterally-varying multi-layer structures. This has become possible as a result of two principal developments. First, a new generation of inversion techniques has been developed and become widely used (Meissner, 1986). These permit modelling not only of travel times, but also of wave forms in the observed P- and S-wave data. Secondly, multichannel data acquisition methods and experimental configurations are now being used which give trace spacings of 100 m or less in crustal wide-angle seismic experiments.

There are several advantages to obtaining such closely-spaced traces. First, it is much easier to pick first breaks, simply because the character and position of an arrival vary only slightly from trace to trace. For the same reason later arrivals can be identified reliably and correlated across a record section from sub-critical reflections to wide-angle reflections and refractions. All these phases then can be used to constrain the model when the data are inverted. In addition to improving travel-time information, closely spaced traces also allow the variation in signal amplitude to be monitored across a record section without spatial aliasing. This too is a valu-

able parameter when modelling the data. A further advantage of collecting this type of wide-angle seismic data is that multichannel techniques can be used in data processing. These include tau-p methods (Diebold and Stoffa, 1981), slant stacking and the application of normal moveout corrections to study reflections. All of these factors help to improve the resolution of wide-angle seismic experiments.

Wide-Angle Data Acquisition Techniques on the Continental Shelf

On land it has been relatively easy to improve the resolution of wide-angle experiments by increasing the number of receivers. This approach presents problems at sea because it is not practical to deploy a large number of fixed recording instruments at close quarters. Similarly, the cost of detonating explosive charges at 100-m intervals would be prohibitive and the operation would be beyond the capabilities of many navigation systems. As a result, data from wide-angle experiments at sea still have a typical trace spacing of 2–4 km (e.g. Barton and Wood, 1984).

There are, however, existing marine multichannel wide-angle seismic techniques. One of these, which recently has been used extensively, is two ship, expanding spread profiling (ESP) (Stoffa and Buhl, 1979). Here two ships steam apart on opposite headings at constant speed. The seismic source is fired from one ship and seismic arrivals are recorded by the other, which tows a multichannel streamer (Fig. 1). The resulting dataset is densely sampled, and from it an accurate, one-dimensional velocity–depth curve can be determined below the mid-point of the line. However, if the crust beneath the line is not laterally homogeneous, the interpretation becomes ambiguous, since the line is not reversed. Two-dimensional structure cannot be resolved because for an ESP, unlike conventional reversed wide-angle experiments, there are no ray paths crossing at low angles. The ESP technique also presents particular problems with static corrections if there are variable thicknesses of sediment below both the source and receiver, neither of which is fixed (Calvert, 1985).

ESP has proved most successful in studies of the oceanic crust, where shot-receiver offsets of 50 km or less are sufficient for observing upper mantle refractions as first arrivals. On the continental shelf, where

a)

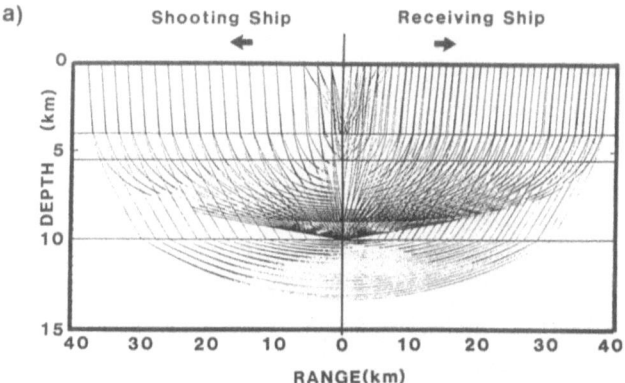

TWO SHIP EXPANDING SPREAD PROFILE

b)

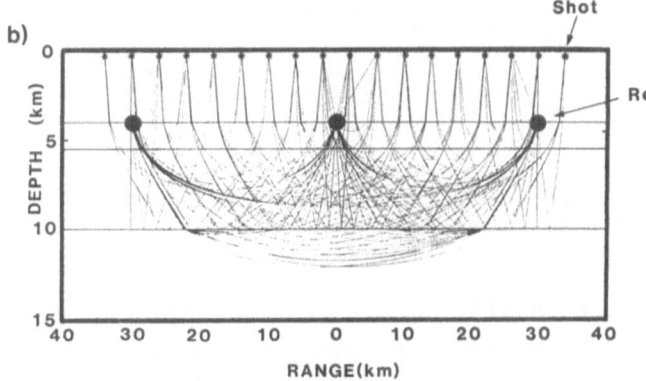

REVERSED REFRACTION PROFILE

Fig. 1. Comparison of the geometries of (a) expanding spread profiles (ESPs) and (b) reversed wide-angle experiments using fixed receivers. Two-dimensional structure can be resolved with the latter configuration, where there are many rays crossing at low angles, but not with ESPs, which contain no crossing ray-paths.

the crust is much thicker, offsets of 120 km or more are needed to see mantle first arrivals. At these ranges, repetitive airgun array sources are not sufficiently powerful to provide a detectable signal. It is also difficult to prepare and fire large explosive shots (ideally 100 kg or more) spaced half a receiver array length (~ 1200 m) apart while steaming at a steady speed of around 6 knots.

The problem of recording closely spaced wide-angle datasets on the continental shelf has been overcome by developing a fixed, multichannel seabottom receiver (Powell *et al.*, 1986). The Pull-Up Multichannel Array (PUMA) was designed with the specific objective of measuring seismic velocities in the lower crust and upper mantle with a greater resolution than had previously been possible in reversed refraction experiments recorded at sea, but

without the limitations of ESP techniques. The system is an array, and therefore offers all the processing advantages of data recorded by a multichannel streamer, but avoids some of the statics problems associated with ESPs because the PUMA is a fixed receiver. Large explosive shots can also be recorded at long range because the firing programme is not restricted by logistical problems as it is for an ESP; for example, there is no need to fire all the explosive shots during a single pass of the shooting line. The paths taken by rays travelling from a given shot to each hydrophone will be different. When the data are plotted in record section, it is assumed that lateral variability at depth does not occur on a scale smaller than the length of the recording array.

The PUMA is a 1.2 km-long, 12-channel hydrophone array and recording system which can be deployed on the seabed in water depths of up to 200 m (Fig. 2). It is suitable for use in reversed, wide-angle experiments, in which airgun or explosive shots are fired at spacings of not more than one array length, to provide a final record section with trace spacings of 100 m or less from normal incidence out to refracted mantle first arrival ranges.

Instrument Description

The PUMA's seismic sensors consist of 12 short hydrophone sections containing piezo-electric transducers in an oil-filled PVC sleeve (Fig. 3). Hydrophone sections are separated by 100 m-long passive array sections. Each is an oil-filled PVC tube containing a steel strength member and conductors which carry power supply to the array and seismic signals back to the recording package. When the PUMA is assembled, adjacent hydrophone and array section terminations are pushed together and sealed by '0 rings', leaving air-filled chambers in between for multiway connectors and hydrophone pre-amplifiers. An alloy sleeve is bolted across hydrophone sections, which are weaker than the interconnecting cables, to maintain the strength of the array along its length (Fig. 3). The sleeve contains machined slots to ensure good acoustic coupling between the hydrophone and the surrounding sea water.

Eleven of the passive sections are identical and therefore interchangeable. In case of damage at sea, a faulty section could be removed on deck and the PUMA redeployed as an 11-channel system. The

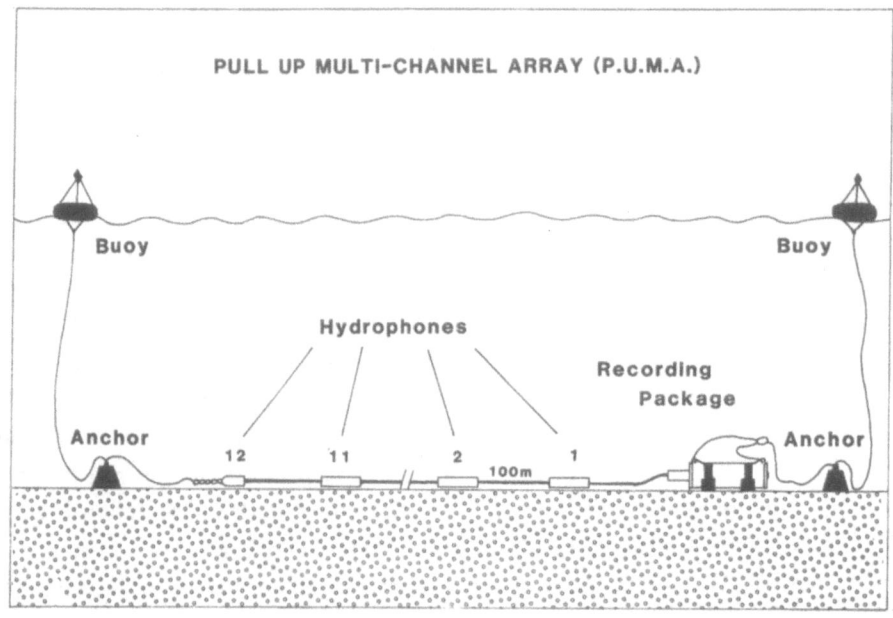

Fig. 2. A sketch showing the PUMA system deployed on the seabed. Twelve long passive sections separate the twelve hydrophones and carry seismic signals to the recording package. The instrument is moored to buoys for recovery.

twelfth passive section connects the hydrophone array to the recording package.

The recording package is contained within a cylindrical pressure vessel with flat end caps (Fig. 4). One end cap carries the connections to the array, while the other is fitted with a watertight bulkhead connector. This can be used for external monitoring and resetting of the PUMA's internal clock. The recording package includes a battery power supply and a set of eight cassette tape recorders as well as recording and control electronics. The tape recorders have

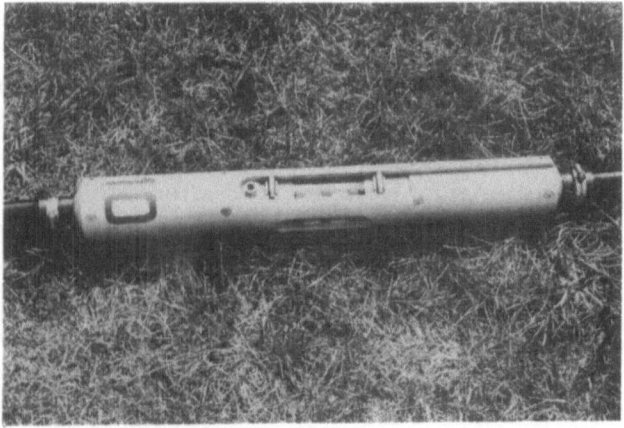

Fig. 3. The assembled hydrophone unit connected between two passive array sections. The external slotted sleeve maintains mechanical strength across the hydrophone module.

been modified for four-track recording. Each is capable of recording seven channels of frequency modulated (FM) analogue data plus clock and flutter correction channels. One cassette recorder can therefore record from half the array (6 channels) while the seventh data channel records one hydrophone at lower gain. Cassette tape recorders are run in pairs with odd and even numbered channels being recorded on odd and even numbered recorders respectively.

Recording electronics filter, amplify and frequency modulate incoming signals and record them on to tape. This method of recording mixed FM signals on cassette tape was implemented on earlier seismic instruments built at Cambridge (Duschenes et al., 1985; Smith and Christie, 1977; Owen and Mason, 1982). The recorded signals are separated and demodulated by a separate replay system, which also produces a signal for digitizing from the flutter channel.

Control electronics monitor, control and activate the recording system, including the tape recorders. They comprise an internal clock, microprocessor controller and relay drivers for switching the tape recorders and recording electronics. The clock is based on a crystal oscillator from which all timing signals are derived. Its output is recorded on to each cassette tape for timing, and its drift calculated by

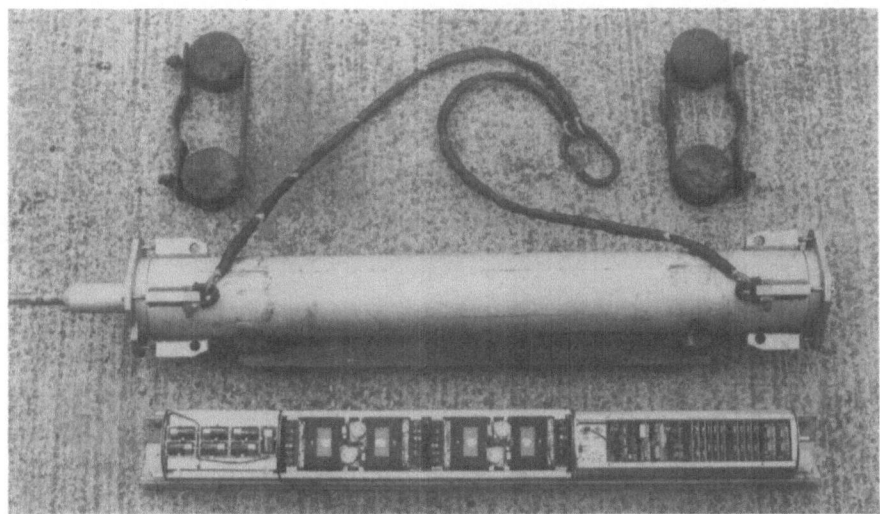

Fig. 4. The recording package (foreground) consists of a chassis holding (from left to right) a battery power supply, 8 cassette tape recorders and the instrument control panel and electronic circuitry. This is housed in a pressure case, which is attached at one end to the PUMA array (left). The pressure vessel is sealed by removable end-caps bolted on at either end.

comparing it with a shipboard clock before deployment and after recovery. The microprocessor unit (MPU), which derives its timing from second pulses from the crystal controlled clock, remotely controls the state of the PUMA. It is responsible for switching the instrument on and off and for activating the correct pair of tape recorders for each predetermined recording window. A table of recording windows may either be downloaded to the MPU directly, or from a disk file on a separate microcomputer. The system allows for up to 128 recording windows to be programmed, each of an integral number of 5-second periods, starting on any second. The MPU will also perform repeated instructions of up to 60 identical on and off cycles. A full description of the PUMA with system specifications is given in Powell et al., 1986; and Powell, 1986 (a) and (b).

PUMA Deployment

The PUMA is deployed from a standard hydrophone array winch mounted on the stern of a research ship (Fig. 5). Its mooring system consists of two sets of buoys, mooring lines, swivels and anchor weights (Fig. 6), so that the instrument may be recovered from either end, depending on tide and weather conditions. The length of cable between the buoy and anchor weight is twice the water depth. The length of wire linking the PUMA to the anchor is also greater than the water depth, so that the anchor

Fig. 5. The 1200-m PUMA on an array winch mounted on the stern of RRS Challenger. The ends of each hydrophone section are supported on the curved drum by wooden blocks. The end of the array (here protected by a dummy end) is attached to the recording package before deployment. One set of mooring wires is wound on to the winch drum underneath the PUMA.

weight reaches the sea bottom before deployment of the PUMA itself begins. The anchors and ground lines decouple the instrument from any noise generated by the buoys and their moorings. Swivels are used on all the mooring wires, so that any torque

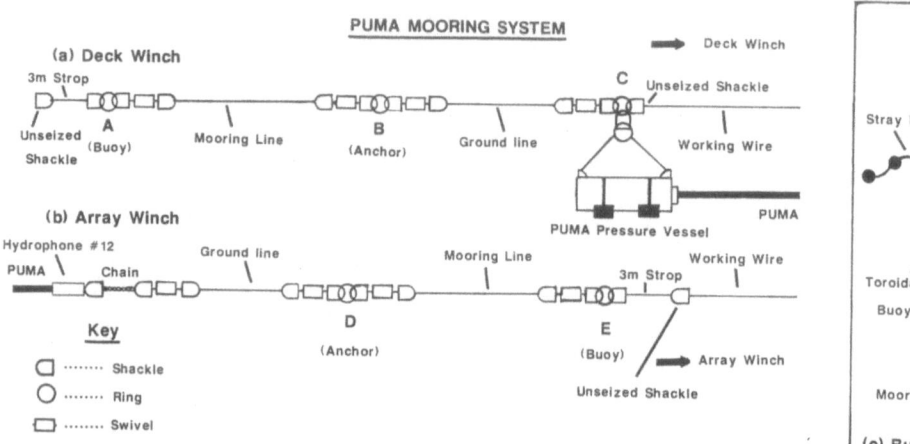

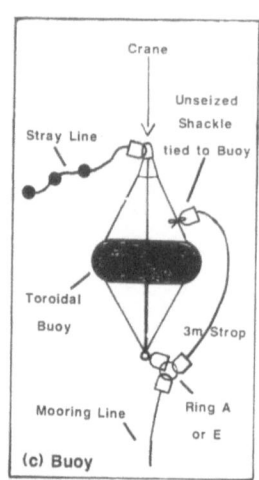

Fig. 6. The PUMA Mooring System. Deployment. (a) Firstly, the toroidal buoy attached at ring A is deployed, followed by a (350-m) mooring wire from the deck winch, the first anchor weight (Ring B), a (250-m) ground line, also from the deck winch, and the PUMA pressure case (Ring C). The end of the PUMA is led out via a protecting chute and attached to the recording package. (b) Once the PUMA has been laid on the seabed a second set of mooring lines, wound on to the array winch underneath the PUMA, are deployed. These begin with a short length of chain shackled to hydrophone 12, and continue with the second (250-m) ground line, anchor weight (Ring D), (350-m) mooring wire and buoy (Ring E). Figures in brackets indicate cable lengths if water depth is 200 m. Recovery. (c) One of the two buoys is first retrieved. The shackle at the end of the 3-m strop is then attached to the main warp, the buoy unfastened from ring A (or E) and the rest of the mooring system and PUMA recovered in reverse order to deployment.

produced in them under tension is relieved at the swivels rather than the less robust PUMA.

The weighted pressure vessel is attached to one ground line via a two-legged strop (Figs 4 and 6). The array and recording package are connected together on deck before being deployed.

Deployment and recovery each begin at either buoy in the mooring system, and then proceed sequentially, finishing at the second set of moorings. However, it is usual to deploy the recording package end first (Fig. 6), and recover it last. The PUMA is laid in a line by advancing the ship at low speed in the desired direction while unwinding the PUMA from the winch drum and through a protecting chute on the after-deck into the water. The PUMA is negatively buoyant and therefore sinks to the sea floor. During deployment the array is kept slightly taut and directly astern to avoid laying it in a non-linear configuration. Launching the PUMA is attempted only in reasonable weather conditions (less than Force 4) and at slack water, to avoid damaging the array.

It is also possible to supply the recording package with new cassettes and batteries and to reprogramme the MPU while leaving the array on the seabed. The pressure case is recovered and detached on deck, and the array temporarily redeployed with a dummy end.

The operation is completed by re-attaching the pressure case to the array and lowering the PUMA back into position.

Experiment and Results

The PUMA was first deployed in a reversed wide-angle experiment west of the Outer Hebrides (Fig. 7; Powell and Sinha, 1987) in summer 1984. The aim of this experiment was to determine the crustal velocity structure with high resolution along the Winch line, a BIRPS deep reflection profile (Brewer et al., 1983).

The PUMA, which was deployed at the northern end of the line, successfully recorded all 99 explosive shots. These were detonated approximately one array length (1100 m) apart at ranges of 18–138 km. At closer ranges an airgun source was used. Other recording instruments included shallow water seismometers (PUSSs; Smith and Christie, 1977) located at A1 and A3, and land seismometers (SCRAPs; Owen and Mason, 1982) on the Isle of Lewis (Fig. 7).

A 5-channel PUMA record section for explosives line A is shown in Fig. 8. Trace amplitudes are scaled for shot size and range, and traces are plotted with a reduction velocity of 7.0 km s^{-1}. The signal to noise ratio is good, and it is easy to follow phases across

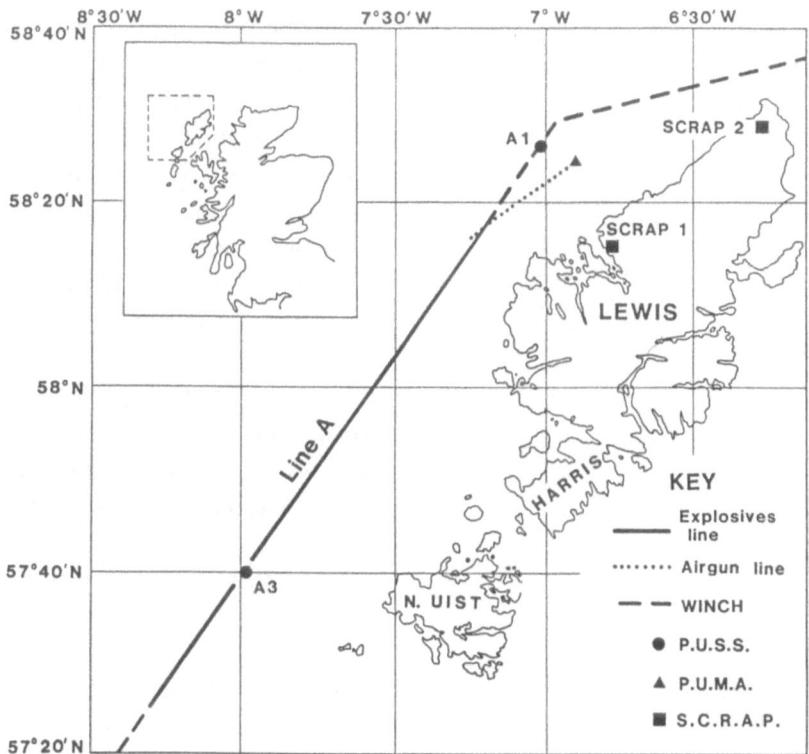

Fig. 7. Schematic representation of the PUMA Experiment, west of Lewis. Explosive line A is coincident with a section of the BIRPS WINCH profile. Shots were recorded by sea-bottom instruments at A1 and A3 and by land-seismometers on Lewis.

the section and to monitor changes in amplitude with offset. The amplitude and travel times of arrivals can be determined with greater accuracy from the multi-channel dataset than from the other single channel data because of the similarity and close proximity of adjacent traces. It is also obvious where the shape of

a first arrival is altered by noise, by comparison with nearby traces.

The first arrival can be identified along the whole record section. It reaches a maximum amplitude at 45–65 km range. At around 85 km its energy decreases sharply, and the first arrival would be

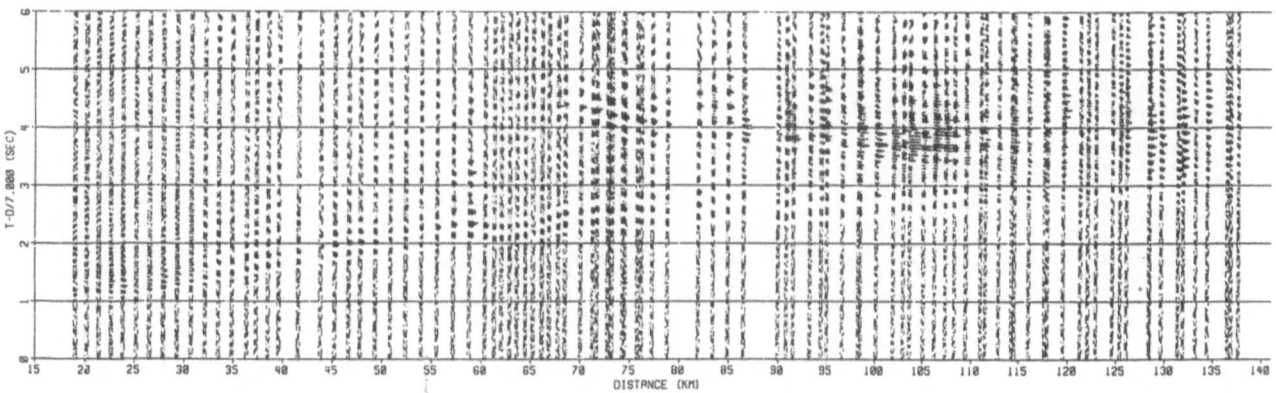

Fig. 8. A 5-channel PUMA record section from explosive line A reduced at 7 km s^{-1}. The traces are plotted with amplitudes scaled for range and shot size, and have been band-pass filtered between 2 and 40 Hz. Gaps in the record section occur because the PUMA was not aligned parallel to the explosives line during the experiment, and because four shots misfired. Note the good signal-to-noise ratio, and the ease with which arrivals can be traced across the record section. The two most prominent phases are the first arrival from the upper and mid-crustal layers, and the wide-angle reflection from the Moho.

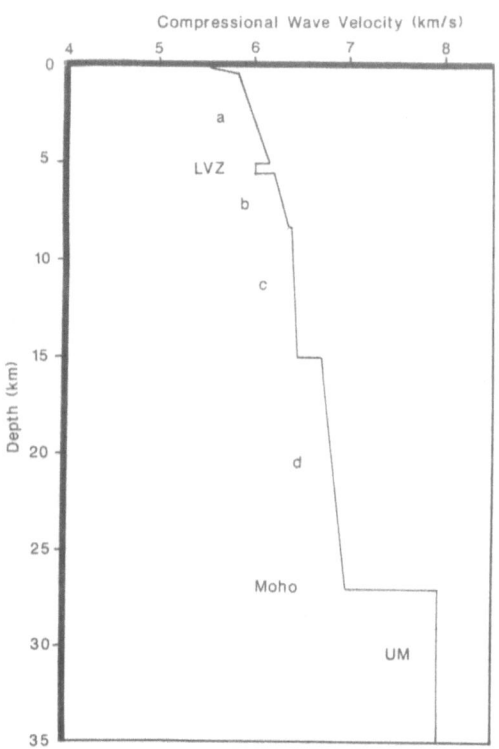

Fig. 9. Velocity-depth curve derived from 2-dimensional amplitude modelling of explosives line data using asymptotic ray theory. LVZ = low velocity zone; UM = upper mantle.

65 km, and between here and 110 km both its duration and its frequency content increase. Its maximum amplitude occurs at 95 km. The Moho refracted phase, *Pn*, is small in amplitude. It does not occur as a first arrival on the PUMA record section, but does on the data from the land seismometers on Lewis. Shear wave phases can also be clearly identified on PUMA record sections. Particularly distinct on the records are high amplitude shear wave reflections from the Moho and diving waves from within the crystalline crust.

Travel times and amplitudes of *P*-wave data have been modelled by ray-tracing (Červený *et al.*, 1977) and using the reflectivity method (Fuchs and Müller 1981). The model which best fits the data is displayed as a 1-D velocity depth curve in Fig. 9, and converted to two-way vertical travel time and superimposed on a line drawing of the BIRPS reflection profile in Fig. 10.

The results of both experiments show much similarity; however, the Moho, which on deep reflection profiles is often considered to lie at the base of the reflective lower crust, appears to coincide with the top of a distinctive reflective zone at 8.3 s two-way travel time. Amplitude modelling of PUMA data does suggest that this boundary is structured (Powell, 1986b).

It is also significant that the PUMA recorded the first wide-angle dataset on the Hebridean shelf from which lower crustal arrivals (layers 'c' and 'd', Fig.

difficult to identify beyond here without such closely spaced traces. Amplitudes increase again at around 125 km offset. The wide-angle Moho reflection can be seen beyond 50 km range reduced at a two-way travel time of about 4 s. Its amplitude increases at

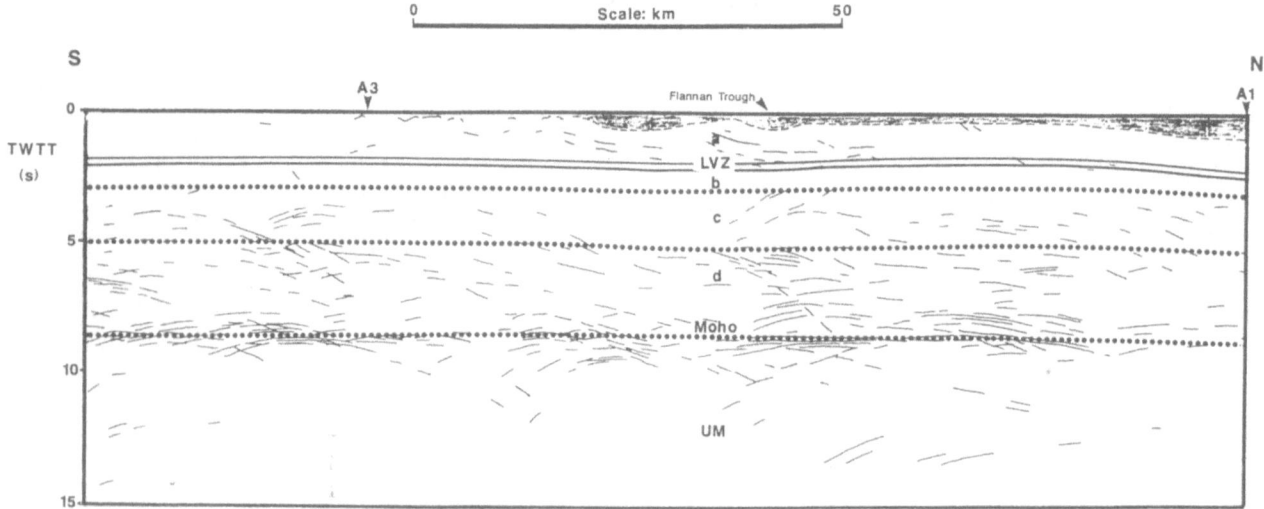

Fig. 10. Comparison of the BIRPS WINCH and PUMA experiment data recorded at wide angle. A two-dimensional model composed of the velocity-depth curve in Fig. 9 for the Flannan Trough has been converted to two-way travel time and superimposed on a line drawing of the WINCH data. Dotted lines denote interfaces. The low velocity zone (LVZ) is indicated by a continuous line. The Flannan Trough is shaded. Annotation as in Fig. 9. The Moho coincides with the top of the banded reflections at 8.3 s two-way travel time.

10) have been successfully identified and modelled (e.g. Jones *et al.*, 1984; Bott *et al.*, 1979). This can be attributed to the improved resolution of low amplitude arrivals from the multichannel dataset.

Conclusions

It has been shown that a new approach was needed to recording seismic data at wide angle on the continental shelf, in order to study the velocity structure of the continental lithosphere in greater detail. A sea bottom multichannel hydrophone array was designed and built to meet this need. The success of the PUMA instrumentation is illustrated by its performance at sea, where it recorded 100% of the data from explosive shots detonated on the sea floor. The quality of this dataset is extremely good. Noise levels are low and, for much of the data, almost every 'wiggle' on the seismograms can be traced across a single shot gather, and many phases are easily correlated between adjacent shots.

Indications are that, providing the instrument is continually modified to keep step with advances in technology, the PUMA, with its unique survey geometry will remain a vital tool for recording multichannel data on the continental shelf.

Acknowledgements

We wish to thank the master, officers and crew of *RRS Discovery* and the staff of *NERC Research Vessel Services* for their assistance at sea. A great many people at *Bullard Laboratories* were involved in various stages of this project, and their contributions are gratefully acknowledged. The work was financially supported by the *Natural Environment Research Council*, Grant GR3/5200. University of Cambridge, Department of Earth Sciences contribution number ES 1249.

References

Barton, P. J., Matthews, D., Hall, J., and Warner, M., 1984, Moho beneath the North Sea Compared on Normal Incidence and Wide-Angle Seismic Records, *Nature (London)* **308**, 55–57.

Barton, P. J., and Wood, R., 1984, Tectonic Evolution of the North Sea Basin: Crustal Stretching and Subsidence, *Geophys. J. R. Astron Soc.* **79**, 987–1022.

Bott, M. H. P., Armour, A. R., Himsworth, E. M., Murphy, T., and Wylie, G., 1979, An Explosion Seismology Investigation of the Continental Margin West of the Hebrides, Scotland, at 58° N, *Tectonophysics* **59**, 217–231.

Brewer, J. A., Matthews, D. H., Warner, M. R., Hall, J., Smythe, D. K., and Whittington, R. J., 1983, BIRPS Deep Seismic Reflection Studies of the British Caledonides, *Nature (London)* **305**, 206–210.

Calvert, A. J., 1985, *Seismic Studies of the Atlantic Fracture Zones: Charlie-Gibbs and Tydeman*, Unpublished PhD Thesis, University of Cambridge.

Červený, V., Molotkov, I. A., and Psencik, I., 1977, *Ray Method in Seismology*, Univ. Karlova, Praha, Czechoslovakia, 214 pp.

Diebold, J. B. and Stoffa, P. L., 1981, The Traveltime Equation, tau-p Mapping and Inversion of Common Midpoint Data, *Geophysics* **46**, 238–254.

Duschenes, J., Potts, C. G., and Rayner, M., 1985, Cambridge Deep Ocean Geophone, *Mar. Geophys. Res.* **7**, 455–466.

Fuchs, K. and Müller, G., 1971, Computation of Synthetic Seismograms with the Reflectivity Method and Comparison with Observations, *Geophys. J. R. Astron. Soc.* **23**, 417–433.

Jones, E. J. W., White, R. S., Hughes, V. J., Matthews, D. H., and Clayton, B. R., 1984, Crustal Structure of the Continental Shelf off Northwest Britain from Two-Ship Seismic Experiments, *Geophys.* **49**, 1605–1621.

Meissner, R., 1986, *The Continental Crust: A Geophysical Approach*. International Geophys. Series, 34, (W. L. Donn, ed.), *Academic Press, London* 423 pp.

Meissner, R., Luschen, E., and Fluh, E. R., 1983, Studies of the Continental Crust by Near-Vertical Reflection Methods: A Review, *Phys. of the Earth and Planetary Interiors* **31**, 363–376.

Mooney, W. D. and Brocher, T. M., 1986, *Coincident Seismic Reflection/Refraction Studies of the Continental Lithosphere: A Global Review*, Presented at the Second International Symposium on Deep Seismic Reflection Profiles of the Continental Lithosphere 15–17 July 1986, Cambridge, England, 42 pp.

Mueller, S., 1977, A New Model of the Continental Crust, in *The Earth's Crust*, Geophysical Monograph 20, American Geophysical Union, 289–317 (J. G. Heacock, ed.), Washington, DC.

Owen, T. R. E. and Mason, M., 1982, *A Cassette Recorder for Explosion Seismology*, Internal Report, Bullard Laboratories, University of Cambridge.

Phinney, R. A., 1978, Interpretation of Reflection Seismic Images of the Lower Continental Crust, *EOS., Trans. AGU* **59**, 389.

Powell, C. M. R., 1986a, *The PUMA Manual*, Internal Report, Bullard Laboratories, University of Cambridge.

Powell, C. M. R., 1986b, *A Wide-angle Multichannel Seismic Study of the Continental Lithosphere*, Unpublished PhD Thesis, University of Cambridge.

Powell, C. M. R. and Sinha, M. C., 1987, The PUMA Experiment West of Lewis, UK, *Geophys. J. R. Astr. Soc.*, **89**, 259–264.

Powell, C. M. R., Sinha, M. C., Carter, P. W., and Leonard, J. R., 1986, A Sea-Bottom Multichannel Hydrophone Array, *Mar. Geophys. Res.* **8**, 277–292.

Smith, W. A. and Christie, P. A. F., 1977, A Pull-Up Shallow Water Seismometer, *Mar. Geophys. Res.* **3**, 235–250.

Stoffa, P. L., and Buhl, P., 1979, Two-Ship Multichannel Seismic Experiments for Deep Crustal Studies: Expanded Spread and Constant Offset Profiles, *J. Geophys. Res.* **84**, 7645–7660.

An Active Source Electromagnetic Sounding System for Marine Use

M. C. SINHA, P. D. PATEL, M. J. UNSWORTH, T. R. E. OWEN and M. R. G. MACCORMACK
University of Cambridge, Department of Earth Sciences Laboratories, Madingley Road, Cambridge CB3 0EZ, UK

(Received 27 April, 1989; accepted 1 September, 1989)

Key words: electrical conductivity, instrumentation.

Abstract. Instrumentation has been developed for carrying out active source electromagnetic sounding experiments in the deep oceans. Experiments of this type are directly and uniquely sensitive to the presence of molten or partially molten material, to temperature structure and to the porosity of upper crustal rocks such as those that accommodate hydrothermal circulation systems. Electromagnetic sounding experiments therefore represent an extremely desirable addition to the existing range of geophysical techniques for studying geological processes in thermally, hydrothermally or magmatically active regions—for example, at oceanic spreading centres.

The instruments can be operated in regions of rugged, unsedimented sea bottom terrain, and are designed for investigating the distribution of electrical conductivity within the oceanic crust and uppermost mantle. The instrumentation consists of a deep towed, horizontal electric dipole transmitter and a set of free-fall, sea bottom, horizontal electric field recording devices.

The transmitter is a deep-towed instrument, which is provided with power from the towing ship through a conducting cable. The transmitter package is fitted with an integral echo sounder, which allows it to be towed safely a short distance above the seabed. Electromagnetic signals are transmitted from a neutrally-buoyant antenna array, which is streamed behind the deep tow.

The sea bottom receiving instruments each consist of a recoverable package which contains the instrumentation and digital recording system, an acoustic release unit, four low-noise, porous electrodes arranged in two orthogonal, horizontal dipoles, and a disposable bottom weight.

The instruments have been used at sea on three occasions. On their most recent use, active source signals were successfully recorded during an experiment to investigate crustal magmatism and hydrothermal circulation beneath the axis of the East Pacific Rise.

Introduction

In this chapter we describe a set of instrumentation developed to carry out active source electromagnetic sounding experiments in the deep oceans. The primary incentive for doing this is a desire to study *in situ* a range of geological processes that occur in the crust and uppermost mantle at oceanic spreading centres. A property of many such processes (for example, the injection of magma from the mantle to crustal levels; the accumulation of molten or partially crystallized material in crustal magma chambers; the production of new oceanic crust by emplacement and crystallization of this melt; and the activity of hydrothermal circulation systems) is that they are all concerned with the presence of fluid phases within the crust and upper mantle, and the interactions between the fluids and the surrounding rocks.

Geological processes in which the activity of fluids plays a major role are by no means limited to mid ocean ridge environments; they occur commonly in a wide range of geological settings. However, it has become evident in recent years that ridges represent the sites of a unique and complex set of interactions: between tectonic, magmatic and hydrothermal processes; between mantle, crust and hydrosphere; and between geothermal, geochemical, hydrological and biological systems. Many of the processes and interactions involved are understood poorly, if at all, and a number may prove to be of substantial economic significance (for example, the role of ridge hydrothermal systems in the genesis of polymetallic sulphide ore bodies).

The distribution of electrical conductivity within the oceanic crust and upper mantle is very largely determined by the presence, even at low concentrations, of magmatic and hydrothermal fluids, conductivities of which can be several orders of magnitude greater than those of the surrounding host rock; and by temperature. Other physical parameters accessible to geophysical measurements, such as seismic velocities or density, exhibit much weaker dependence on these factors. Experiments which allow reliable determinations of electrical conductivity structure are therefore directly and uniquely sensitive to the presence of fluid phases and temperature anomalies, and are therefore potentially a more fruitful source of structural information about fluid or thermally related processes than other marine geophysical methods. The development of practical, electrical or electromagnetic sounding techniques that can be applied in the deep oceans is therefore a highly desirable goal for the study of these processes, either at mid ocean ridges or elsewhere.

The most commonly used technique for deep electrical sounding on land is the magneto-telluric (MT) method. However, this technique suffers from a serious limitation in the deep oceans. This is that the presence of several kilometres of highly conductive sea water effectively screens the seabed from the ionospheric sources of MT signals at all frequencies higher than a few cycles per hour. The very long period signals that do penetrate to the seabed can be used for large-scale sounding of the sub-oceanic mantle (e.g. Filloux, 1981), but are so severely limited in terms of depth resolution that they are quite unsuitable for intra-crustal investigations.

An alternative approach to MT sounding is to generate artificial source currents in the water column, and to measure the resulting electric or magnetic fields at the seabed. This approach makes it possible to exploit the high conductivity of sea water, both by using it as the return path for the artificial source currents and by taking advantage of the electrically quiet environment at the seabed which its screening properties provide.

One successful example of an artificial source technique for electrical sounding of the upper oceanic crust has been described by Edwards et al. (1981). Their technique, known as MOSES (Magnetometric Off-Shore Electrical Sounding), involves the use of a vertical transmitter array and of magnetometers for measuring the resultant fields at the seabed. It is essentially a galvanic resistivity method, in which a long vertical transmitter with its lower electrode close to the seabed is used to inject current into sub-sea bottom formations. Changes in the magnetic field at the seabed due to reversing the source current are measured by vertical component, sea bottom magnetometers.

A second and entirely different technique, known as active-source electromagnetic sounding, has been described by Chave and Cox (1982), and Young and Cox (1981). In this method, which provides probably the best approach to the problem of how to obtain detailed conductivity information from the deeper parts of the oceanic crust and from the sub-oceanic upper mantle, horizontal electric dipoles are used for both the source and the receivers. The technique is a frequency–domain sounding system, which relies on the variations in the response of the earth to electromagnetic signals of different frequencies for determinations of conductivity structure.

This chapter describes a newly developed instrumentation system for active source electromagnetic sounding of the oceanic lithosphere which is based on the principles described by Chave and Cox, and which has recently been developed at Cambridge University.

Active-Source Electromagnetic Sounding

The instrumentation system developed at Cambridge for electromagnetic sounding consists of two main elements (Fig. 1). The first of these is a Deep-towed Active Source Instrument (DASI). DASI is towed from a surface ship on a conducting cable, and transmits electromagnetic signals at extremely low frequencies (0.0625–256 Hz) into the seabed. The second element is one or more Low-frequency, Electro-Magnetic Underwater Recorders (LEMURs). LEMURs are free-fall instruments which are deployed on the seabed at an appropriate range (typically a few kilometres) from the transmitter, and used to record the resultant oscillating electric fields.

In the active source electromagnetic sounding method, the earth's conductivity structure is determined by measuring its response to radiation over a wide range of frequencies. The deep towed source instrument generates oscillating electric and magnetic fields close to the seabed, by passing a large, alternating electric current between a pair of electrodes mounted in an antenna streamer which is towed horizontally behind the instrument package. The return path between the transmitter electrodes is provided by sea water. Towing the transmitter

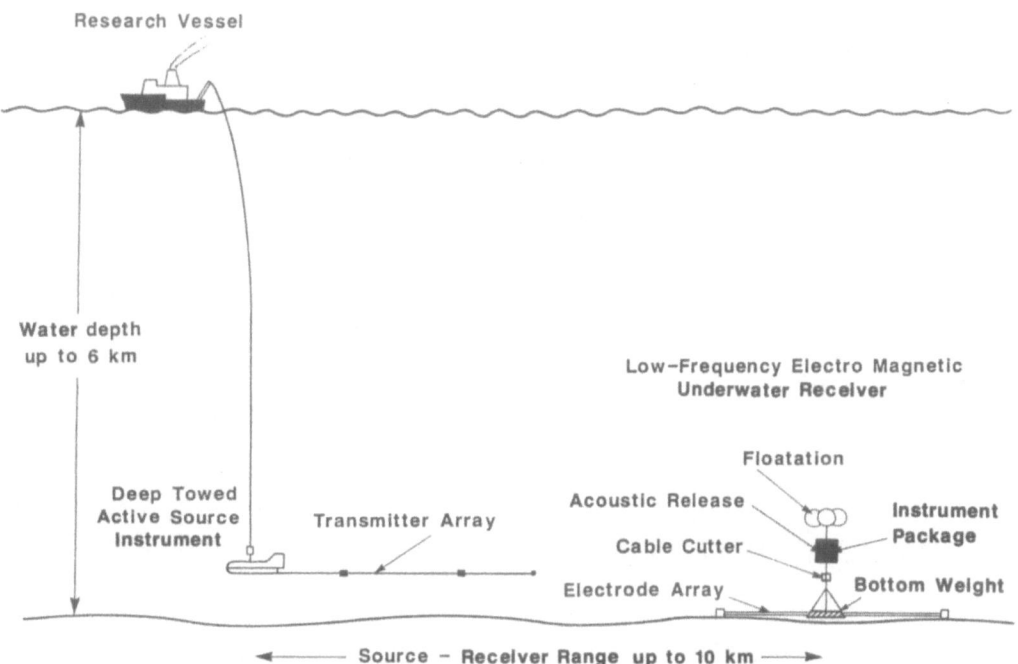

Fig. 1. Schematic arrangement of an active source electromagnetic sounding experiment, showing the deep-towed active source instrument, the surface research vessel and a low-frequency electromagnetic underwater recorder (not to scale).

antenna sufficiently close to the seabed ensures that the electric and magnetic fields are well coupled into the sub-bottom rocks.

Because of the high conductivity of sea water, radiation from the transmitter is very rapidly attenuated in the water column. However, in the relatively resistive seabed, the signal is able to propagate for much greater distances. By placing suitable recording instruments on the seabed at an appropriate distance (typically a few kilometres for oceanic crustal studies), the attenuation and phase shift of the transmitted signals can be measured, and related to sub-bottom conductivity. Since the depth of penetration of the transmitted signal into the seabed depends on its frequency, signals at decreasing frequencies contain information about the conductivity of the earth to increasing depths below the sea floor. Thus by making measurements at suitable ranges and over an appropriate set of frequencies, it is possible to derive information about the variation of sub-bottom conductivity with depth. Use of several receivers and numerous source positions can obviously provide information about lateral, as well as vertical, variations in sub-bottom conductivity.

Deep-Towed Active Source Instrument (DASI)

The deep towed active source instrument, or DASI, constitutes the transmitter end of the Cambridge electromagnetic sounding system (Fig. 2). It consists of three major sub-systems: the ship-board power supply unit, control unit and echo-sounder display (Fig. 3); the deep-towed instrument package (Fig. 4); and the transmitter antenna array. The deep-towed package is connected electrically and mechanically to the ship-board systems by means of an armoured, co-axial cable.

The ship-board power supply unit converts the ship's 3-phase mains supply (440 V, 50 Hz, unstabilized) to a variable, regulated, frequency stabilized supply at up to 2000 V RMS and 5 A RMS. The frequency of the output supply is 256 Hz, and this is tied to a frequency standard derived from the ship's master clock system. This stabilized, high-voltage output is fed to the deep tow cable to power the deep-towed transmitter.

At the deep tow, the power received from the ship is first transformed down to a lower voltage, and then switched by a semiconductor bridge to provide a pattern of half-cycles of either positive or

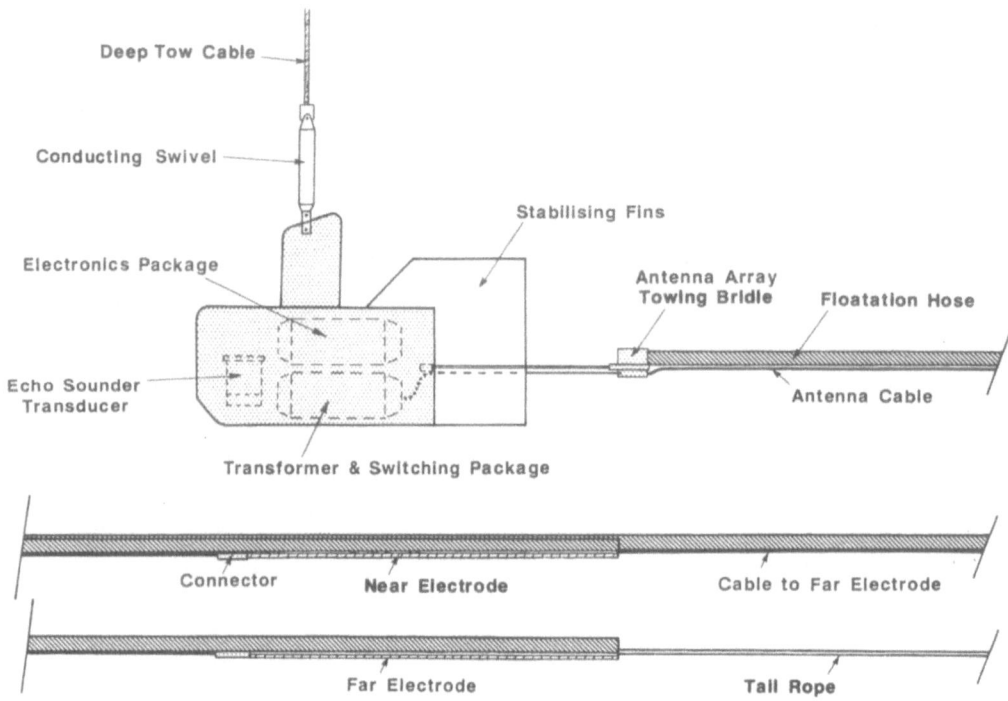

Fig. 2. The deep-towed active source instrument, DASI. The instrument package is suspended from the surface ship by an armoured conducting cable. Two pressure vessels house the active source signal generator, the echo sounder transceiver and the associated control and monitoring equipment. The 180-m long, neutrally buoyant antenna array is streamed horizontally behind the deep-tow package.

negative polarity to the transmitter antenna array. The maximum output current to the antenna is 200 A RMS. By controlling the switching to provide blocks of appropriate numbers of half-cycles of each polarity to the antenna, pseudo-square waves of any desired frequency that is an integer factor of 256 Hz can be synthesized. Obviously, since the transmitted frequency is derived directly by dividing the power supply frequency in this way, the transmitted frequency has the same stability characteristics as the power supply unit and the ship-board clock system.

DASI's height above the seabed is monitored by means of an integral echo-sounder. The frequency of this is selectable at either 3.5 or 7 kHz. The lower frequency has the advantage that it provides significant penetration in thinly sedimented areas. It is therefore possible to determine the instrument's height above igneous basement, as well as its height above the seabed, in such areas. The signals from the echo sounder are frequency modulated on a 50-kHz carrier, and telemetered to the towing ship along the deep-tow cable. Tuned coupling circuits at the ship and deep-tow end enable the carrier signal to be injected into the cable and retrieved from it while it is simultaneously carrying the high voltage, 256 Hz power supply. Data from the echo sounder are demodulated at the towing ship and displayed on a graphic recorder. A winch operator on the ship uses the display to maintain the deep tow as closely as possible at a fixed height above the seabed, by veering or heaving in cable. Since the echo sounder transducer is effectively omni-directional, a reflection from the sea surface is usually seen, as well as that from the seabed. Consequently, the echo sounder indicates the depth of the deep tow beneath the surface, as well as its height above the seabed.

Local control of the deep-tow instrument is performed by a microprocessor unit. The microprocessor controls the switching of the semiconductor bridge in the electromagnetic transmitter, and also the pulse length, frequency and receiver gain of the echo sounder. The microprocessor also performs monitoring and diagnostic tasks via a multiplexed, analogue to digital interface. This allows it to measure the time-averaged voltage and current at the antenna; the instantaneous current and voltage, in a high-frequency sampling mode which logs the wave form of the transmitter and stores it in memory,

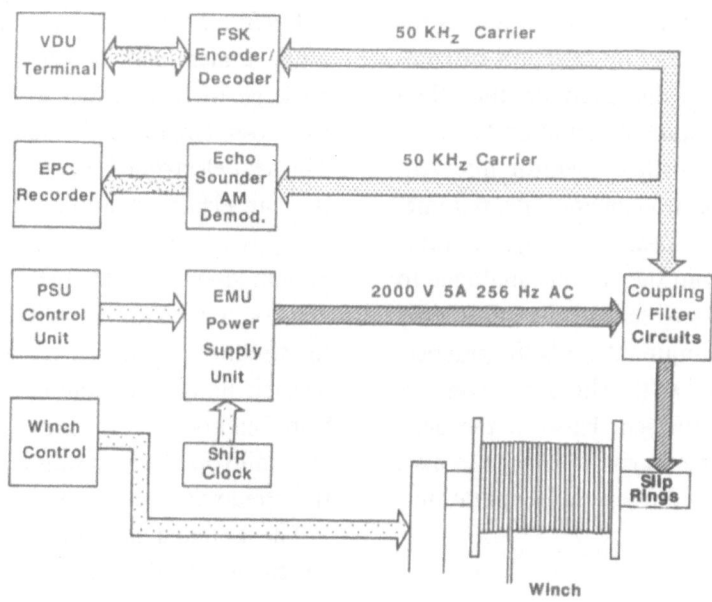

Fig. 3. EMU Transmitter—shipboard systems for the deep-towed transmitter. These systems provide a stabilized, high-voltage power supply to the deep tow; serial communications with its microprocessor controller; and display of data from its 3.5/7.0 kHz echo sounder.

for subsequent transmission to the towing ship; the temperature of the cooling oil in the pressure vessel housing the transmitter bridge and transformer; and the state of the instrument's various battery packs. In addition, a moisture-detection circuit continually tests for high impedance between a pair of electrodes located at the bottom of the oil-filled pressure vessel, in case of seepage of sea water into the system.

The microprocessor in the deep tow communicates with a VDU terminal at the surface ship through the deep tow cable by using a serial communications link running via a frequency-shift keyed (FSK) carrier centred on 50 kHz. The FSK link shares this carrier frequency with the echo sounder data. An additional task of the microprocessor is to organize time sharing of the carrier between outgoing, analogue echo sounder data, outgoing FSK data and incoming FSK commands from the operator's terminal on the ship.

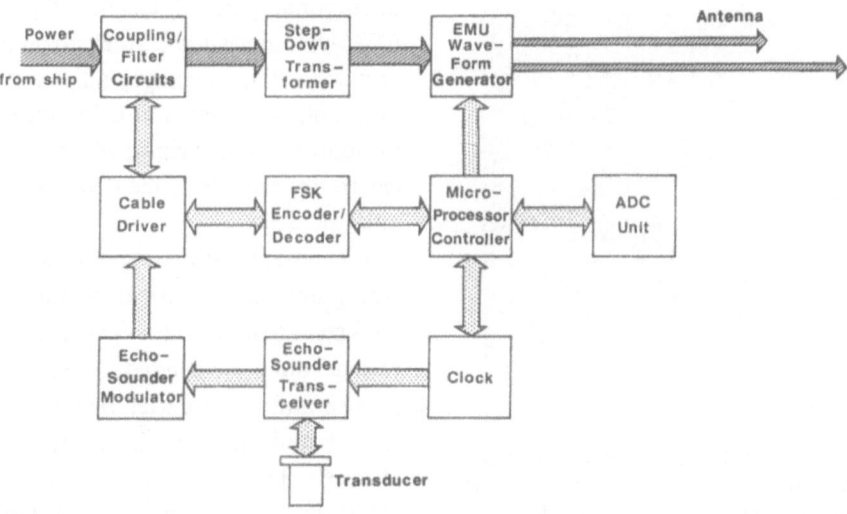

Fig. 4. Block diagram of the deep-towed active source instrument. The step down transformer and wave form generator convert power supplied from the ship and feed it to the antenna array. The remaining electronic systems provide an acoustic indication of the instrument's position in the water column, and control and monitoring functions.

Software running on the microprocessor allows for simple control of the deep tow by selection of command options from menus displayed on the VDU screen. Once a particular mode of operation has been selected and initiated, no further operator intervention is required until one or more operational parameters need changing. In the meantime, regular status reports (including data from the various analogue to digital interface channels) are sent to the console by the microprocessor at 2-minute intervals. In practice, a personal computer is used as the shipboard console, so that data and commands sent between the deep tow and the console can be logged to magnetic disc, providing a permanent record of deep tow operations.

Fig. 5. Photograph of the deep-tow system while it is being deployed. The conducting swivel which terminates the deep-tow cable is positioned immediately above the instrument. The electrode array is visible in the background, suspended from the instrument package and streaming astern of the ship just below the sea surface. The array storage and handling winch can be seen in the right foreground.

DASI's antenna array consists of a pair of heavy-duty, underwater cables connecting the deep-tow package to two electrodes. The electrodes are made from 6-m lengths of 12-mm diameter steel wire rope. The near electrode is towed 30 m behind the deep tow, the far electrode another 100 m behind that. The array is supported by a number of 30-m lengths of 63-mm diameter, flexible PVC hose filled with low density, low compressibility oil. Neutral buoyancy of the streamer is achieved by adjusting the volume of oil in each buoyancy section, and by distributing small lead weights along the array. A 50-m long, polypropylene tail rope terminating in a small drogue provides the streamer with additional towing stability.

The antenna is deployed from a large-drummed winch, identical to those used for handling seismic streamers. Once the array is in the water, its inboard end can be connected electrically and mechanically to the deep-tow package. The entire deep-tow system can then be deployed relatively easily over the stern of the towing ship (Fig. 5). The deep tow cable is terminated by a deep water conducting swivel system, fitted immediately above the deep tow package, to minimize torsional stresses in the cable.

Low-Frequency, Electromagnetic, Underwater Recorder (LEMUR)

The receiving instruments, or LEMURs, are free fall, ocean bottom instrument packages which digitally record changes in horizontal electric field across an orthogonal pair of electric dipole antennas. Each LEMUR consists of: an electronics, instrumentation and recording package; an acoustic release unit; a mechanical release and cable cutter assembly; a disposable bottom weight; a set of disposable, porous electrodes and electrode deploying arms; and a buoyancy package.

When deployed from the ship, the LEMUR is configured as a vertical string, with the buoyancy package at the top, the instrumentation and acoustic release units suspended beneath it, the mechanical release and cable cutter below that, and the bottom weight and electrode array assembly at the bottom (Fig. 6). The instrument sinks to the seabed, where after the bottom weight has landed the electrode deploying arms fold down into a horizontal position, leaving the electrodes in a cross configuration of two orthogonal, 13.5-m dipoles.

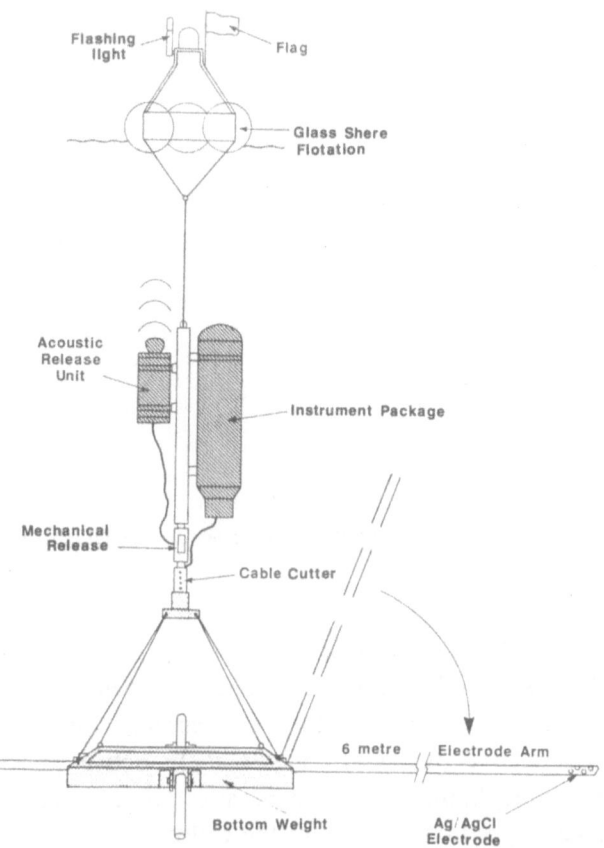

Fig. 6. Schematic arrangement of a low-frequency, electromagnetic underwater recorder (LEMUR). The instrument package and acoustic release unit are mounted on a stainless steel frame and suspended beneath the floatation package. The bottom weight and electrode array assembly are connected to the instrument frame by the mechanical release and cable cutter units.

At the end of an experiment, the instrument is recovered by acoustically activating the release unit. The acoustic release system operates the mechanical release and cable cutter, severing all connections between the instrument package and the bottom weight. The instrument package, still attached to the buoyancy, then floats back to the surface.

The LEMUR's electrical sensors are a set of four low noise, porous, silver–silver chloride electrodes which are deployed in a horizontal array around the bottom weight. The electrodes are based on a design developed at Scripps (Webb *et al.*, 1985). A silver foil is mounted on a cylindrical PVC former, and surrounded by a mixture of *Kieselguhr* (diatomaceous silica) and silver chloride. The mixture is held in place by an external sleeve of porous polyethylene. The electrode is immersed in sodium chloride solu-tion under a partial vacuum to remove air bubbles, and then electrolysed so that a uniform layer of silver chloride forms on the outer surface of the silver foil. This process makes low noise, low impedance elec-trodes which are sufficiently robust to withstand operations in a marine environment. However, once wet and electrolysed, the electrodes have a relatively short storage life of 6 to 8 weeks, so the final stages of preparation are carried out at sea immediately before use.

Each electrode is fitted into the end of a 6-m long deploying arm, made of polypropylene tube. The end of the tube, where the electrode is housed, is pierced by a pattern of holes to allow good electrical contact between the electrode and sea water. The inboard end of each arm is attached by a hinge to the bottom weight. The hinge has a restricted travel, allowing the arm to move in a vertical plane from 10° below horizontal to 70° above horizontal. This movement allows the arms to fold upwards during the instru-ment's descent through the water column. Solid-glass weights placed in the ends of the arms close to the electrodes ensure that once the instrument reaches the bottom, the arms lie flat on the seabed.

The bottom weight itself consists of a rectangular steel frame with sides of 1.4 and 1.56 m weighing 65 kg. A set of rope strops attach it to the bottom of the cable cutter assembly in an arrangement that maintains the instrument package above it in a fixed orientation relative to the electrode array.

The cable cutter is a guillotine, actuated by the mechanical release unit of the acoustic command system. The mechanical release breaks the load-bearing connection between the instrument package and the body of the cutter unit. The cutter then falls rapidly towards the sea floor, driven by a 4-kg weight attached to it. The guillotine blade remains attached to the instrument package by a short length of stainless steel wire. When the cutter reaches the end of the wire's travel, the blade is pulled out of the cutter body, severing the four electrical cables which connect the electrodes to the recording package. The cutter blade returns to the surface with the instru-ment package, but the cutter body remains on the seabed.

The acoustic release system is the standard unit made by the Institute of Oceanographic Sciences, Deacon Laboratory (Phillips, 1981). It operates at a centre frequency of 10 kHz, and makes use of the

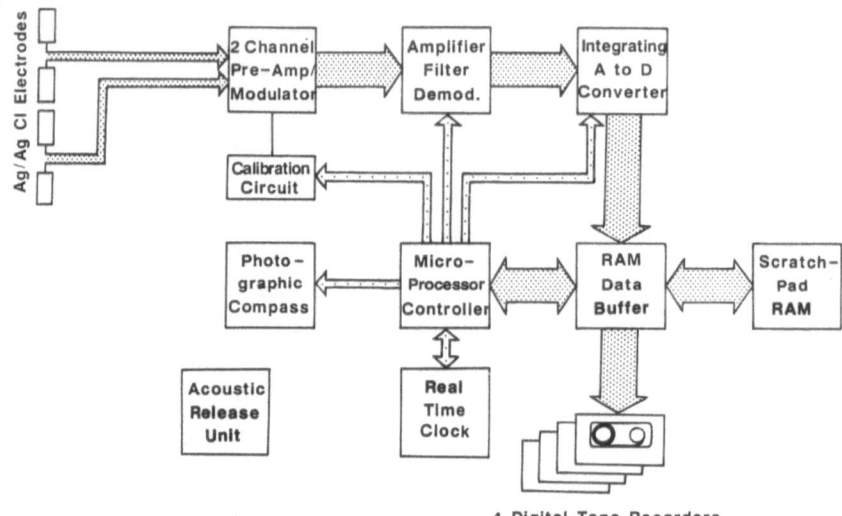

Fig. 7. Low-frequency Underwater Electromagnetic Recorder (LEMUR). Block diagram of the LEMUR electronics and digital recording system (see text for a description).

ship's precision echo sounder system for graphical display of the range and status of the release unit. The instrument's buoyancy is provided by three 0.43-m diameter glass spheres, mounted on an aluminium frame. A mast and flag and a pressure deactivated flashing light are fitted to the buoyancy unit to aid recovery of the instrument by day or night.

The LEMUR instrument package is contained in a cylindrical pressure vessel, capable of operating at water depths of up to 6 km. In the instrument package, voltage signals from two data channels, corresponding to opposite pairs of electrodes, are amplified, frequency filtered, digitized, stacked and digitally recorded on magnetic tape (Fig. 7).

Each electrode is coupled via an electrolytic capacitor to one side of a differential preamplifier. At the preamplifier, the voltage between electrodes is chopped by a field effect transistor bridge, and then coupled through a step-up transformer to a high-gain amplifier which is tuned to the chopping frequency. After amplification, the chopped signal is demodulated, frequency filtered, amplified further and then digitized. The analogue to digital converter is based on a voltage-controlled oscillator and counter-timer. This forms a 16 bit, time-integrating digitizing system, which has better signal-to-noise characteristics than sample-and-hold types of converter.

The digitized signals from both channels are stored in random access memory, and subsequently transferred to magnetic tape. The digital tape system uses four cassette drives, totalling approximately 20 MBytes of data storage. The data can be synchronously stacked as they are recorded. Since in general, the frequency content of the active source signal is known in advance, selection of suitable stacking parameters allows the collection of a great deal more data than would otherwise be possible, without significantly degrading the data quality. Automatic gain control software dynamically adjusts the gains of the amplifiers, to maintain maximum dynamic range of recording without saturating the analogue to digital converters.

All aspects of the recording system, including the sampling rate and resolution of the analogue to digital converters, the fold of stack, the automatic gain control and the cut-off frequencies of the high and low pass filters, are under the control of a microprocessor unit. Data are recorded on tape in discrete recording windows, each representing 16 kilobytes of data from each channel. Data tables containing the start times and recording parameters for up to 620 windows are stored in random access memory. The microprocessor continually scans these tables, and initiates recording with the appropriate settings at the beginning of each window. All of the parameters controlled by the microprocessor and mentioned above can vary from window to window, allowing great versatility in the instrument's overall recording programme. The data tables can either be prepared directly on the microprocessor, or more conveniently they can be prepared in advance on a personal computer and downloaded to the instrument through its serial communications port.

Calibration circuits are built into the preamplifier stage, and can be activated by the microprocessor. The calibration circuits allow step functions of voltage or current to be applied individually to either sensor channel. In this way, the calibration circuits provide measurements of the sensor impedance and of cross-talk between channels, as well as of the system gain and sensitivity.

Orientation of the instrument is achieved by a photographically recording compass system which is located in the instrument package. A photographic image of a compass needle is recorded on film during data recording. Since the orientation of the recording package is fixed relative to the bottom weight, this allows the orientation of the receiver electrode array relative to magnetic north to be determined.

The design of the LEMURs draws to a large extent on that of the Cambridge Digital Ocean Bottom Seismometer (Owen and Barton, 1989). The recording package pressure case, the cable cutter, the microprocessor unit, the digital tape recording system and the buoyancy package make use of the same, or identical, hardware. Use of this existing technology has made possible the development of a sophisticated sea bottom instrument within the resources of a relatively small university research group in a short period of time.

Experience of Use

To date, the equipment described here has been used at sea three times. The first occasion was in September 1977, during an instrument test cruise in the north Atlantic. Experience on that cruise brought to light a number of problems, among the more serious of which was the difficulty encountered in telemetering echo sounder and serial digital data through the deep-tow cable while simultaneously using it to supply power to the electromagnetic transmitter. However, the cruise did confirm that the equipment was being developed along broadly correct lines.

The instruments were used for a second time, after modifications and improvements, in July and August 1988. This was during a cruise to the Lau Basin, a back-arc, marginal basin in the western Pacific. The objective was to carry out a combined programme of seismic and electromagnetic sounding studies of the Valu Fa Ridge, a back-arc spreading centre located in the southeastern part of the basin. Again, a number of instrumental problems affected the electromagnetic sounding equipment, and we were not able to collect an interpretable electromagnetic data set. The cruise did, however, act as a useful proving ground for the modifications made to the instruments since the first cruise, and provided experience of operating them over the difficult terrain of a spreading centre.

During the 1988 cruise, we were able to operate the transmitter close to the seabed at antenna currents of up to 120 A RMS for a total of 16 hours. Electric field data collected by LEMURs deployed on the Valu Fa Ridge indicated RMS noise levels as low as 10^{-7}–10^{-8} V m^{-1} in the frequency band 0.0625–8 Hz. The noise is dominated by low frequencies, and represents a combination of environmental background noise, which predominates at the lowest frequencies, and electrode noise, which predominates at higher frequencies. The levels are consistent with measurements of sea bottom electric fields reported by Webb et al. (1985).

The combination of sustainable transmitter power and low receiver noise levels achieved during the cruise gave us confidence that, provided the remaining instrumental problems could be overcome, the equipment would be capable of generating data that would provide useful information about crustal conductivity structure beneath an active ridge system.

After further changes, the equipment was used at sea for a third time on the East Pacific Rise in May and June 1989. Improvements to the instrumentation for this cruise included modifications of the LEMUR bottom weight and electrode arm assemblies, to allow them to be deployed more reliably on areas of unsedimented, ridge axis sea floor.

During the cruise, an electromagnetic sounding experiment was successfully carried out on the axis of the East Pacific Rise at 13°15′ N. The experiment was performed in collaboration with the active source electromagnetic sounding group at Scripps Institution of Oceanography. Four Cambridge LEMURs and four similar receiving instruments from SIO were deployed in two 10-km long lines, one along the ridge axis and one parallel to it but offset by 5 km to the east. The Cambridge deep towed transmitter was towed repeatedly along each line in turn for a total of two and a half days, transmitting signals at between 0.125 and 8 Hz.

Having at last solved the problem of telemetering echo sounder signals up the deep-tow cable while sending power down it, we were able to maintain DASI's height above the seabed at 30 ± 10 m. With the aid of acoustic navigation, it proved possible to manoeuvre DASI along the desired tracks to within 50 m laterally for much of the time, and to within 200 m for virtually all of the time.

At the time of writing, the data from the 1989 cruise have yet to be processed and interpreted. However, the experiment represents the first successful electromagnetic sounding of an oceanic spreading centre, and as such its significance lies not only in the results that will accrue from it but also in the demonstration that such experiments are possible.

Conclusions

The electromagnetic sounding instrumentation described in this chapter represents one of a very limited number of systems capable of exploring the conductivity structure of the sub-oceanic crust and uppermost mantle. The development of a deep-towed transmitter that can be towed reliably a short distance above the seabed, and which uses a neutrally buoyant streamer for its antenna, means that for the first time it is possible to generate high power, low frequency, electromagnetic signals over unsedimented crust in areas of rugged sea bottom topography, such as exist at mid ocean ridges. Our experience and that of other groups indicates that it is possible to exploit such signals successfully for investigating fluid-controlled processes in the seabed beneath the deep oceans.

This chapter cannot be a final or definitive description of the Cambridge electromagnetic sounding system since the instruments are under continual development. However, our aim has been to demonstrate that it has proved possible to construct instruments suitable for electromagnetic sounding of tectonically active, unsedimentented areas—where technical limitations have previously prevented such work from being carried out. The potential scientific and commercial benefits from such investigations are enormous, and it seems likely that the development and use of electromagnetic sounding systems of vari-

ous types will play an increasingly important role in future geophysical investigations of the sub-sea-bottom lithosphere.

Acknowledgements

We are indebted to the members of the active-source sounding group at Scripps Institution of Oceanography, and especially to Chip Cox, Steve Constable and Tom Deaton, for their encouragement of our endeavours, for much useful advice, and for making freely available to us many details of their instrumentation—much of which we have simply copied. We also acknowledge the contributions of John Leonard, Peter Carter, Melvyn Mason and Roger Theobald, who carried out much of the work on instrument development and construction; the officers and crews of the research vessels Discovery and Charles Darwin, and the staff of NERC Research Vessel Services, Barry, who made possible the work at sea; and Phyl Fisher, who prepared the figures. The work was financially supported by the Natural Environment Research Council, under grants GR3/5851, GR3/6671 and GR3/6673.

University of Cambridge, Department of Earth Sciences contribution number ES 1399.

References

Chave, A. D. and Cox, C. S., 1982, Controlled Electromagnetic Sources for Measuring Electrical Conductivity beneath the Oceans—1, Forward Problem and Model Study, *J. Geophys. Res.* **87**, 5327–5338.

Edwards, R. N., Law, L. K., and DeLaurier, J. M., 1981, On Measuring the Electrical Conductivity of the Oceanic Crust by a Modified Magnetometric Resistivity Method, *J. Geophys. Res.* **86**, 11,609–11,615.

Filloux, J. H., 1981, Magnetotelluric Exploration of the North Pacific: Progress Report and Preliminary Soundings near a Spreading Ridge, *Phys. Earth Planet Inter.* **25**, 187–195.

Owen, T. R. E. and Barton, P. J., 1989, The Cambridge Digital Seismic Recorder for Land and Marine Use, *Tectonophys.* (in press).

Phillips, G. R. J., 1981, *The I. O. S. Acoustic Command and Monitoring System*, Report No. 96, Institute of Oceanographic Sciences (Deacon Laboratory), Wormley, Surrey.

Webb, S. C., Constable, S. C., Cox, C. S., and Deaton, T. K., 1985, A Sea Floor Electric Field Instrument, *J. Geomag. Geoelectr.* **37**, 1115–1129.

Young, P. D. and Cox, C. S., 1981, Electromagnetic Active Source Sounding near the East Pacific Rise, *Geophys. Res. Lett.* **8**, 1043–1046.

Long-Range Underwater Photography in the Deep Ocean

Q. HUGGETT

Institute of Oceanographic Sciences, Deacon Laboratory, Brook Road, Wormley, Godalming, Surrey, GU8 5UB, UK

(Received 27 April, 1989; accepted 1 September, 1989)

Key words: Photography, cameras, television, film, surveying, visibility, light sources.

Abstract. Although the optical properties of seawater at extreme depths are more suitable for underwater photography than those at the surface or on continental shelves, they still impose severe limitations on 'long-range' wide area bottom photography. Additionally, deep ocean operations impose technical limitations on control, power and bandwidth. This chapter reviews the approaches contemplated or made towards improving the camera-to-target range in underwater photography in the deep ocean. Further significant improvements await advances in control, power/light sources and bandwidth reduction. With the developments now contemplated, TV and video systems will eventually present a strong challenge to emulsion film techniques.

Introduction

Since William Thompson's production of the first underwater photograph in 1856 (Brown, 1985) several technological advances have made underwater photography an important tool for use in ocean surveying. Rebikoff (1967) presents a succinct history of underwater photography and traces the development of underwater photography from Louis Boutan's early work with glass plate cameras enclosed in brass and iron pressure cases through to the wide-angle lenses of the 1960s. A useful bibliography on advances in underwater photography up to 1968 was published by the Kodak company (Anon. 1968).

As early as 1899, when Louis Boutan lowered his camera system to a depth of 50 m, he had identified the four main hurdles to underwater photography:

- Scattering of light by sea water;
- Data retrieval;
- Need for pressure-resistant housings;
- Provision of adequate power sources.

The first two of these problems have yet to be overcome and continue to plague us today.

Until the mid 1970s underwater photography in the deep ocean was carried out mostly at close range to the sea floor (Heezen and Hollister, 1971). This was done either in a random "pogo-stick" fashion (Laughton, 1957), in stereo (e.g. Ohta, 1983), using a "pop-up" time-lapse system (Lampitt and Burnham, 1983), from a submersible (Ballard and Moore, 1977) or using a sledge which remains in contact with the sea floor (Aldred *et al.*, 1976; Chardy *et al.*, 1980) (Fig. 1). Though even at such close ranges good sharp photographs can be obtained, they cannot cover sufficient ground to give a truly representative picture of the sea floor. These systems are particularly useful for detailed studies of the benthos (Rice *et al.*, 1979; Rice and Collins, 1985) and in the identification of small objects seen in long-range photographs (Fig. 2). However, if we wish to map some of the larger-scale benthic processes, wide-angle or long-range photography is essential.

The desire to view larger areas of the deep ocean floor has led us into two principal areas of difficulty in underwater photography, those of light scattering and data retrieval. These two topics will be discussed with reference to recent developments in deep ocean underwater photography.

Light Scattering

Since the early days of underwater photography the problems of light absorption and scattering in seawater have been paramount. In the absence of ambient (natural) light, and at short ranges from the target, colour balancing can be achieved by using broad band light sources and suitable filters. This was amply illustrated by Mertens (1970), who compared underwater photographs, illuminated by daylight, taken with and without colour filters and at

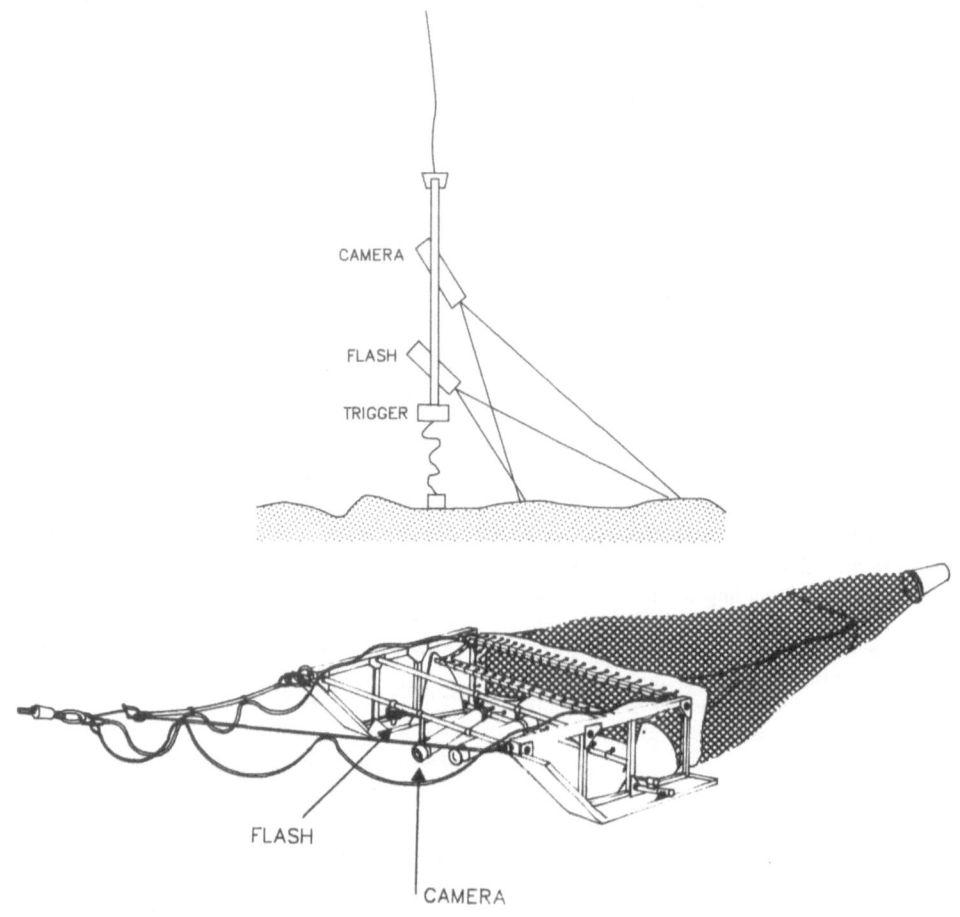

Fig. 1. Illustrations of (A) a 'Pogo-Stick' camera, and (B) the I.O.S. Epibenthic Sledge (after Aldred *et al.*, 1976).

different depths. At longer camera-to-subject ranges (i.e. >5 m) light absorption by the seawater renders much of the visible waveband unuseable for underwater photography. Figure 3 illustrates the effective absorption distances for the various colour bands. Not only is much of the visible waveband strongly attenuated and redundant as a source of illumination, but it can also scatter close to the camera. Thus light which does not reach the sea floor can still contribute backscattered noise to the image.

Over the years, two main approaches have been adopted to ameliorate the effects of backscattering. Both involve the geometric separation of source and receiver; they are:

1. Horizontal camera-to-light separation;
2. *Li*ght *Be*hind *C*amera (LIBEC).

Both these approaches have the advantage of cost effectiveness as all the components in the system are readily available (see Anon. 1986; Anon. 1987) and

are well tried and tested. They have also produced a wealth of good-quality photographs from the world's oceans (e.g. Ballard and Moore, 1977; Phillips *et al.*, 1979; Huggett, 1987). In order to further enhance image definition or camera-to-target ranges, increasingly complex (and expensive!) techniques can be used to reduce backscatter; these are:

3. Light polarization;
4. Monochromatic light sources;
5. Range gating;
6. Synchronous scanning.

These six techniques will be described in order of relative cost and complexity and their merits discussed.

HORIZONTAL SEPARATION

This approach takes advantage of the attenuation properties of the sea by increasing the distance between camera and light source to at least 2 m. This

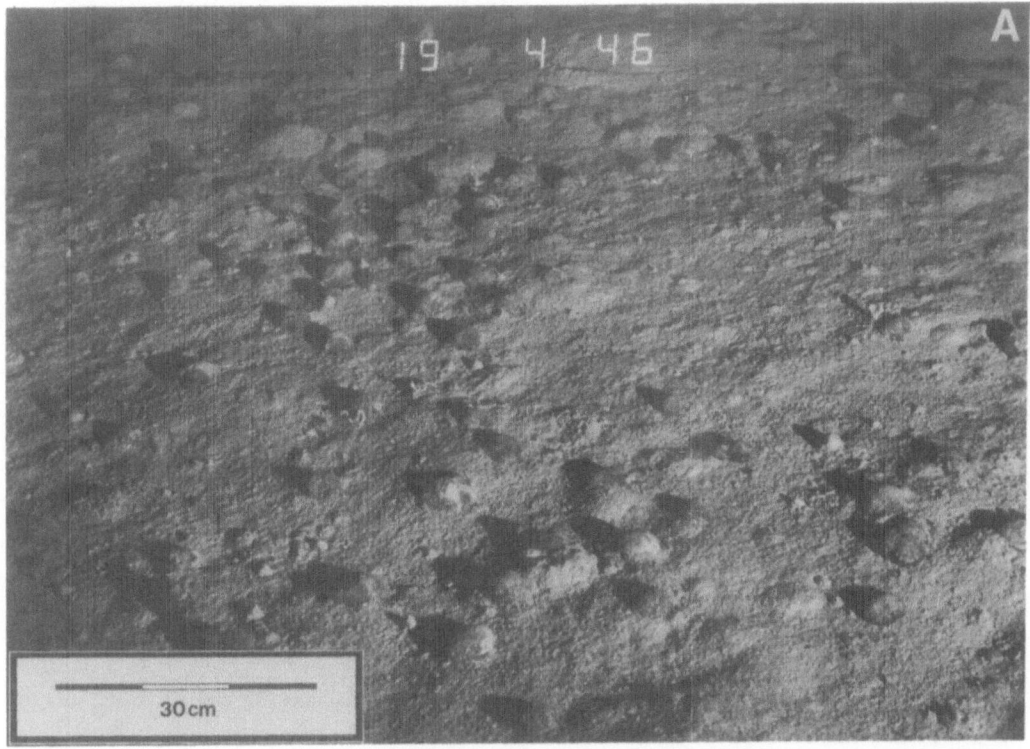

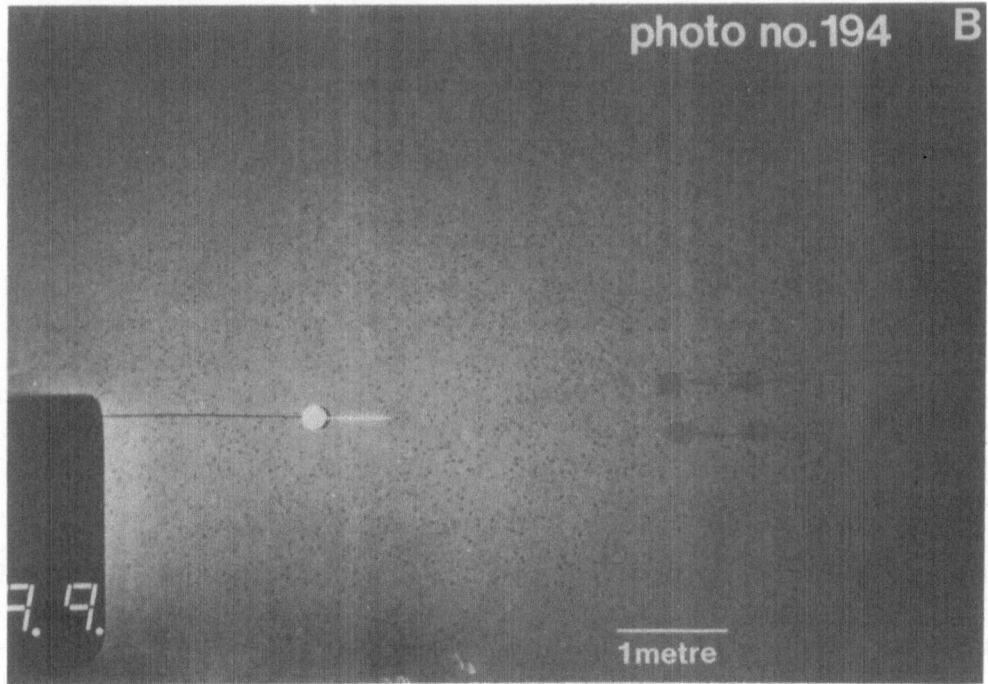

Fig. 2. Two underwater photographs of the same field of manganese nodules taken from (A) the Epibenthic Sledge and (B) the Wide Area Survey Photography (WASP) system to illustrate their differing resolution and scales.

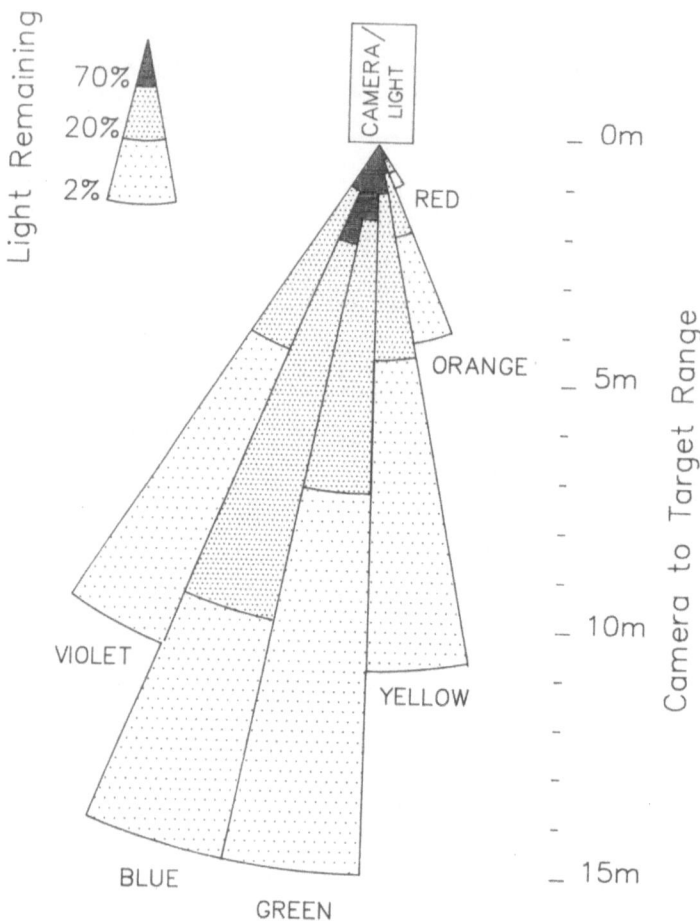

Fig. 3. Diagram illustrating the absorption of light in water. The absorption has been calculated for two-way travel. This illustrates the 'transmission window' in the blue-green waveband.

moves the volume of water common to both light and camera away from the camera so discriminating against the highly attenuated wavelengths which produce negligible image. Although this technique does not eliminate all of the backscattered light, at least some of the wavebands which only produce noise are eliminated by the increased light-to-camera separation.

Figure 4 illustrates the Wide Area Survey Photography (WASP) system developed at the Institute of Oceanographic Sciences (UK) which is typical of systems using horizontal separation. It was designed along the lines of the ANGUS system (Phillips *et al.*, 1979) and uses a still camera with a 37 mm lens and has a 1600 frame capacity. Light comes from 6 Vivitar flashguns powered by NiCad battery stacks capable of at least 2000 discharges per run. A high-frequency echo sounder measures the height of the

camera from the sea floor every 2 s. This information is sent both to the camera, where it is printed on the appropriate negative, and to the ship (via a 10 kHz pinger), for the winch operator to control the camera height. The system is programmed to take photographs when it is within 8 and 18 m of the sea floor.

As an indication of the popularity of this approach to backscatter reduction, several camera systems of this type exist, for example:

- *The ANGUS system*, operated by Woods Hole Oceanographic Institution (Phillips *et al.*, 1979).
- *The RAIE system*, operated by CNEXO (Mauviel, 1982).
- *The USGS Camera system*, operated by USGS, Menlo Park. (Chezar and Lee, 1985).

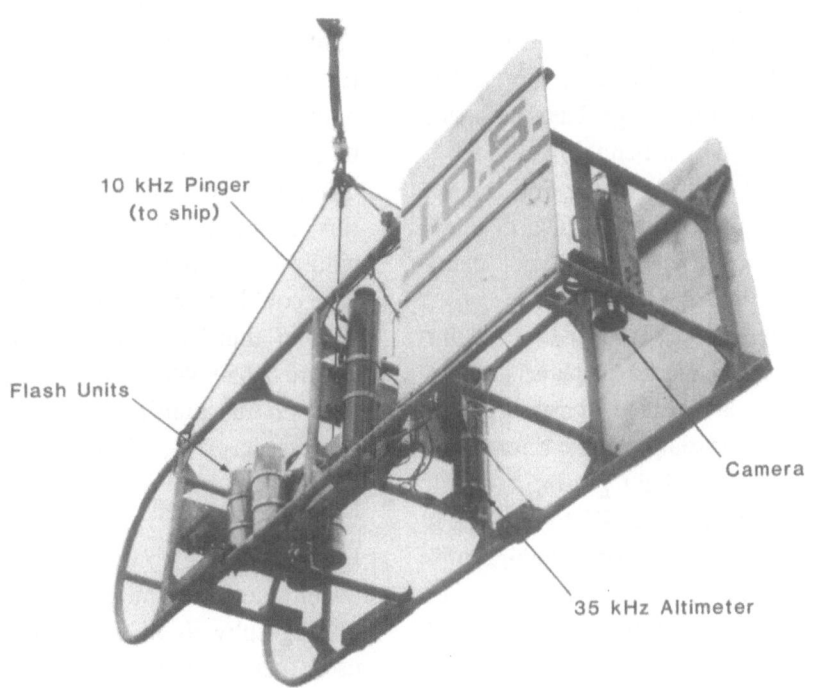

Fig. 4. Wide Area Survey Photography (WASP) system.

In addition to still camera, the USGS Camera system has a video camera and recorder.
- *The ARGO system*, operated by Woods Hole Oceanographic Institution (Harris and Ballard, 1987). This has a TV system in addition to the still camera.

LIBEC

This was developed at the US Naval Research Laboratory (Patterson 1972). Figure 5 illustrates the camera-to-light geometry of the system. This set-up was developed because the introduction of wide-angle lenses increased the volume of water common to both light and camera, so increasing image deterioration. Again, this system separates the light and camera in order to reduce the common volume, so reducing backscatter. Of the geometric separation techniques, this one produces the greatest camera-to-target ranges (up to 21 m; Patterson, 1972). The only (minor) disadvantages of this system are its greater vulnerability to damage and cost of light sources.

LIGHT POLARIZATION

This idea came from experiments using polarized light to see through fog (Nathan, 1957). It involves a plane polarizing filter placed in front of the light

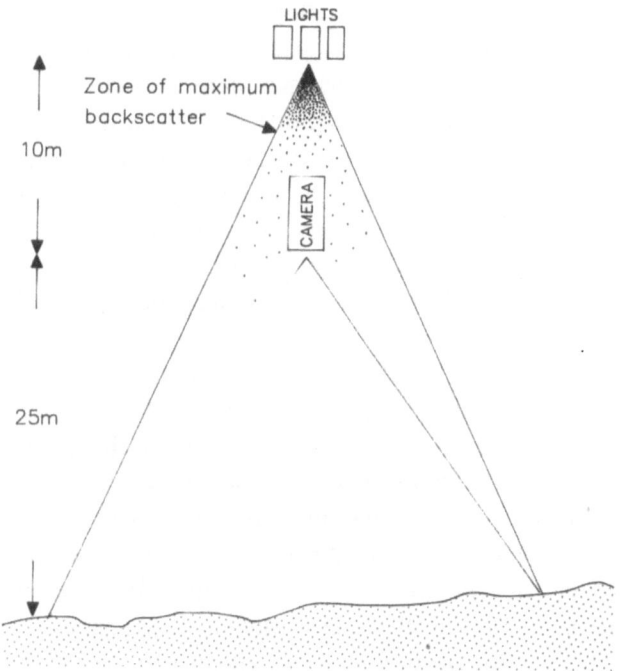

Fig. 5. LIBEC geometry (after Patterson, 1975).

source and a correspondingly crossed polarizer in front of the camera. It works on the principle that the sea-floor can depolarize light, whereas scatterers in the intervening water do not. Thus the camera, with its crossed polarizer, can only receive light from the sea floor. Experiments carried out by Briggs and Hatchett (1965) produced a 20% increase in the visible range in muddy water. The problem with this technique, however, is that it depends on the depolarizing characteristics of both the water and the sea floor, neither of which are absolute. It also depends upon the accurate alignment of the two polarizing filters.

Gilbert and Pernicka (1966) improved on this technique by using circular instead of plane polarizers. Circular polarization is obtained by passing light through a quarter wave retardation plate after it has been linearly polarized. If the light passing through the retardation plate is polarized at 45° to its axes, it emerges with the two optic axes having equal amplitude but with a phase lag of 90° between them. In this condition the light is said to be "circularly" polarized and can propagate with a right- or left-handedness, depending upon which way round the optic axes were orientated in the retardation plate.

The principle behind this approach is similar to that of the plane polarizing technique, as it depends upon the different polarizing characteristics of the sea floor and suspended particles. This is because whenever circularly polarized light is reflected it changes its polarity (or handedness). Suspended particles are small and only produce single reflections and therefore opposite polarity light (to the source). On the other hand, the sea floor produces multiple reflections which result in equal amounts of both opposite and normally (or left- and right-handed) polarized light.

The advantages of this system are that the polarizing characteristics of the sea floor and suspended particles match the circular more than the plane polarizing technique. Circular polarizers are also better because the orientation of the filters is not critical. Cocking (1976) pointed out that the effectiveness of circular polarization depends upon the size of particles causing the backscattering. It appears that circular polarization is most effective in reducing scattering from particles smaller than 1 micron. With particles greater than 6 microns circular polarization can actually make matters worse. Generally, 90% of particles suspended in oceanic waters within 50 m of the sea floor are smaller than 0.8 μm (microns) and 99% of particles are smaller than 6 μm (W. R. Simpson, pers. comm., 1988).

The disadvantage of using polarizers is that light levels are significantly reduced. With a reduction of 50% for each filter crossing, only 25% of the light remains for film exposure. Thus extra expense in film type and/or light source is incurred when using this type of system. For the WASP system, this technique was deemed too expensive as it would have required replacing all the lights. The film type could not be changed since in the UK, only one type of thin-based film is available in long lengths (Ilford X650; 400ASA; HP5 process) and a compromise on the number of exposures per run was not acceptable.

MONOCHROMATIC LIGHT SOURCES

Figure 3 shows how much of the visible spectrum is unsuitable for illumination in long-range underwater photography. If only the optimum wavelengths are used then backscatter will be significantly reduced. There are three ways in which "monochromatic" (narrow-band) light may be produced for underwater photography. These are by filtering, and the use of arc lamps and lasers.

Filtering is best achieved by using a filter which mimics the attenuation characteristics of the anticipated camera-to-target range. The *Wratten Filter Book* (Anon, 1953) gives technical data for Kodak filters and may be used to choose the filter most suitable for the water and range conditions anticipated. Mertens (1970) compared pure water at four target ranges (5, 10, 50 and 100 m) with the four most suitable filters. In turbid waters allowances for a shift of the "transmission window" towards the yellow part of the spectrum must be made. The disadvantage of filtering is that it reduces the overall light level, so that more, or stronger, light sources are needed. The advantage of this system is that it can be adapted to suit different water and range conditions.

Arc-lamps have been built to produce dominant outputs in the blue-green light band. They are based upon the ionization of a mixture of mercury and thallium under high pressure in a quartz tube. Harford (1968) tested this light source and found that the dominant emissions occur at a wavelength of 535 nm (blue-green) and that it has an output efficiency 80 lm W^{-1} (lumens/watt) (i.e. much higher than incandescent tubes which produce up to 40 lm W^{-1}). The thallium arc-lamp is the most suitable arc-light source for long-range photography (Hittleman and Strickland, 1968). It combines efficiency with monochromatic output in the "transmission window" of seawater. The disadvantage of arc-lamps is that they must be operated continuously (for all practical purposes). They therefore need to be cable- rather than battery-powered.

A wide range of lasers of differing spectral output have been produced for underwater application. The most suitable as a monochromatic light source is the copper chloride laser, used in conjunction with a range-gated imaging system (Dixon *et al.*, 1984). Although this laser has not been used as a light source for non-gated cameras its spectral distribution matches the 'transmission window' of seawater and in this respect would be ideal for long-range photography. The main disadvantage is that the cost and complexity of a laser may not be justified for conventional photography alone. Also, the relatively low power output of this source requires an image intensifier to detect the sea floor images.

RANGE-GATING

In the LIBEC system described earlier (Fig. 5), the signal-to-noise ratio (i.e. the ratio of image to backscattered light) was improved by moving the camera closer to the target. In range-gating the effect is similar except that the camera is moved in time rather than space. This is done by generating a short pulse of light and opening the camera shutter only when the pulse is expected to return from the target and for a time equal to the duration of the original pulse (Heckman, 1966). Figure 6 illustrates this technique and shows how most of the backscattered light is prevented from entering the camera.

Two range-gated systems are under development at present; Dixon *et al.* (1984) describe a range-gated system using a pulsed copper chloride laser with a photomultiplier tube (PMT) as the receiver. A laser is required in order to produce a short enough pulse to eliminate unwanted scattering (i.e. a pulse length 10 ns long = 2 m water path length). In this system the laser is pulsed and scans the sea floor. The signal is received in a PMT and converted into raster format. From tests with their prototype system, Dixon *et al.* (1984) envisage a system capable of being operated at 50 m altitude covering an area 100 m wide with a 20-cm resolution. Wilson (1986) described a range-gated system that is likely to become commercially available in the near future. This system uses an image-intensifier camera and a yttrium-aluminium-garnet (YAG) laser. The YAG laser (which operates at a wavelength of 1.06 μm) is fitted with a frequency doubler to produce light at 532 nm, and a beam-spreader in order to match it to the camera viewing angle. So far this

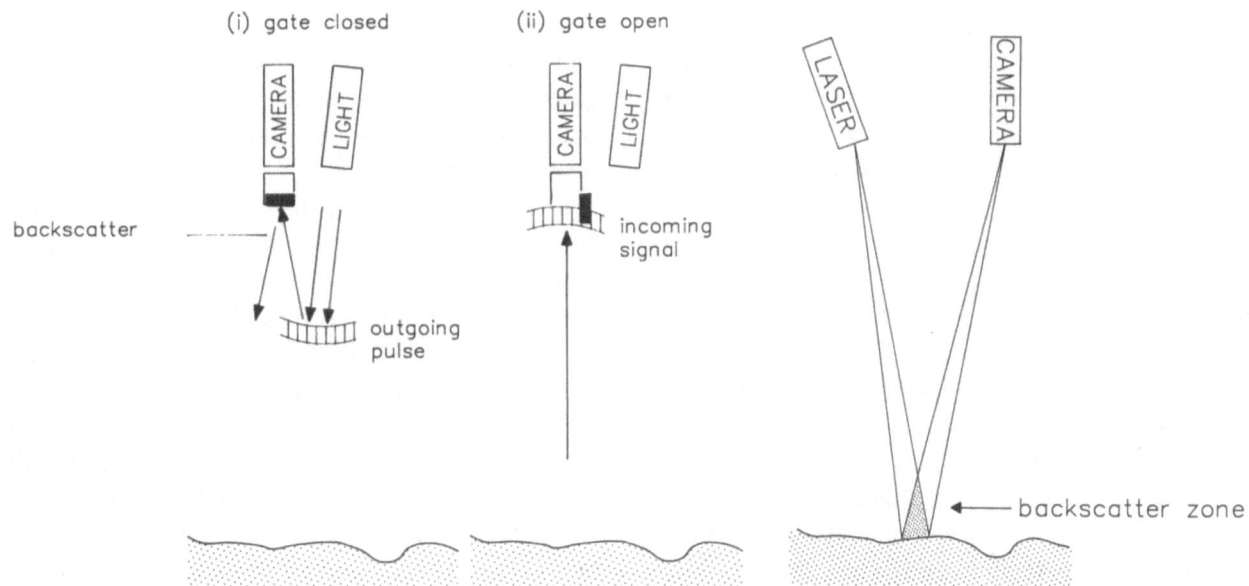

Fig. 6. Diagrams illustrating (A) Range-Gating and (B) Angle-Gating.

system has been tested successfully over a maximum viewing range of 40 m. In principle, range gating should be capable of much greater ranges than this. However, the beam is spread to illuminate the whole viewing area at once. The power per unit area of the laser is reduced as a result and this is the limiting factor at present. For comparison, were it possible for the lamp to be pulsed, a thallium arc would produce 6 J per pulse at 50 Hz compared to 25 mJ for the frequency-doubled, YAG laser. The narrow beam width of the Dixon *et al.* (1984) system puts more power per unit area on to the target (by up to 5×10^4) and is less power-limited.

Range-gating gives underwater photography a leap ahead in survey coverage from the previous record of 21 m altitude set by LIBEC to the 50 m proposed above (corresponding to a sixfold increase in the area covered). There are limitations to this technique however, the main ones being target resolution and receiver sensitivity. All optical images are degraded by forward scattering and refraction from the target. Duntley (1963) found that in macroscopically uniform water, the apparent contrast of fine details decreased inversely as the cube of the subject distance. The "flickering" of a far away object is an example of this on land. In water, forward scattering is more severe and the resulting loss in contrast can limit underwater photography according to the size and type of target to be identified. In designing an underwater viewing system the type of target to be identified must be considered first. This is because target shape and colour will define the contrast conditions that will enable its identification. For this reason a simple rule of thumb for system design is difficult to establish. A good example of this is given by Ryan and Rabushka (1985), who used a camera system (incorporating horizontal separation) from altitudes up to 30 m to search for the Titanic. High-contrast targets such as this are exceptional in underwater photography so altitudes quoted in this paper are based upon natural (low contrast) sea floor targets. Duntley (1963) has written a comprehensive account of the problems of image contrast and describes some field experiments to investigate them. Receiver sensitivity can limit the maximum operating range because pulsed laser sources give a relatively low power output. However, image intensifiers are readily available (Anon, 1986) and their usefulness is

illustrated by Funk (1973). It must be only a matter of time before image intensifiers allow range-gated systems to exceed the 50-m maximum range discussed here.

SYNCHRONOUS SCANNING

Synchronous scanning (or angle-gating) removes scattered light by using a narrow beam (<0.1 degree divergence) to scan the target, reflections from which are analysed by a synchronously scanning receiver (Fig. 6) (Wall, 1968). The receiver cannot see light which is being scattered outside the receiver lightpath so large-angle forward scattered light is also eliminated in this system. To optimize scatter reduction the beamwidth must be small, and there should be some separation between source and receiver to reduce the common volume.

Wall (1968) proposed that synchronous scanning should be carried out in a TV format. Therefore the target would be scanned at a speed that requires time as well as angle synchronization between source and receiver. This is because at target ranges in excess of about 5 m, the travel time for light to and from the target is longer than its 'dwell' time at any point. Thus for any target range, a delay between the output and receive scans must be inserted to achieve synchronization. This delay introduces an element of range-gating to the system so further reducing backscatter.

The obvious disadvantage of this system is its complexity, and likely cost. No synchronous scanning systems are available commercially at present and most effort seems to be concentrated on range-gated systems. The main advantages of this system is that it improves contrast through the removal of forward scattering.

It is clear, therefore, that light scattering will always be a big problem in underwater photography. Until range, or angle-gated systems are further developed and become widely available, we are left with a short menu of techniques for long-range photography. These will allow us to photograph the sea floor from a maximum altitude of 21 m (using conventional film and LIBEC geometry) or 40 m (with range-gated video). If man-made objects (e.g. wrecks, etc.) are the targets for underwater photography, LIBEC altitudes can be increased by approximately 25% since image contrasts will be much greater. Range-gated video is power limited at present but may increase in range as this problem is overcome.

An interesting but extremely low-resolution approach to eliminating optical scattering is to adopt a system that uses sound rather than light to illuminate the target. Jones and Gilmour (1976) describe two slightly different sonic cameras operating at 2 MHz and 3 MHz for use in turbid water. A 2 MHz sonic camera was also tested in a variety of conditions by Andrews (1980). He comments, "... with optical visibility of less than 1 cm ... there was no deterioration of the (sonic) image quality ... A comparison with a low-light TV camera was revealing in that even with good natural lighting in the shallow tank none of the objects could be detected". Andrews (1980) envisaged a theoretical resolution of 86 mm at 10 m altitude. In turbid and extreme circumstances the sonic camera could prove to be a useful tool.

Data Retrieval

In 1893 Louis Boutan was only too aware of the problems of data retrieval as he struggled with the reloading of his 5 × 7 in plate camera. Since those days a variety of imaging systems, with increasing sensitivity, have been developed. These range from the first cellulose films of the 1890s (with emulsions capable of up to 30° BS) to the I.S.I.T. cameras of today, capable of up to 200 000 ASA.

In long range photography the type of receiver used controls the sensitivity, resolution, endurance and playback time of the system. These aspects of image quality will be discussed for the two principal types of receiver:

1. Emulsion film cameras;
2. Television.

EMULSION FILM CAMERAS

First invented at the turn of the century, cellulose-based emulsions were first used in underwater photography in about 1915 to produce underwater, motion pictures for the classic film *20 000 Leagues Under the Sea*.

Most underwater cameras use 35 mm cellulose film on to which a great variety of emulsions can be coated. For long-range photography the emulsion must be blue sensitive in order to record images that have passed through the "transmission window"

(Fig. 3). It is a convenient coincidence that most black and white films have a narrow response in the blue/green wavebands. Red-sensitive monochromatic or colour films are not normally used in long-range underwater photography as broad band-emulsion sensitivity is wasted when the image comprises narrow-band light. A possible exception to this rule is proposed by Fritz *et al.* (1972), who describes a two layer emulsion "colour" film tailored to the transmission window. Underwater trials of this film do not appear to have been made however, and its usefulness is unknown. Unfortunately, the very narrow-band blue-sensitive films have very low sensitivities (around 8 ASA) which limits their use.

Typically, cameras are designed with 90 mm (diameter) spools capable of holding up to 30 m of film. Some films are manufactured with an "Estar" base which can be made thinner than normal films. About 60 m of thin-based film can be loaded on to a 90-mm spool, thereby doubling its capacity. If the increased capacity of thin-based film is desired, then film choice is restricted. In the UK only Ilford X650 (HP5) meets the requirements outlined above.

Mechanical cameras cannot be range- or angle-gated, so the maximum altitude from which photographs can be taken using cellulose film is 21 m. Assuming an average camera altitude of 15 m and camera viewing angles of 55 × 35° (i.e. not wide angle), the average picture area is 15 × 10 m. At a towing speed of 1 m/sec (2 kn), and allowing 10% overlap between frames, one survey can cover a strip of sea floor 21.5 km long by 10 m wide in about 6.2 hours. About the same as a 14-mile strip of two-lane highway!

TELEVISION

The commercial development of television for domestic users has provided us with a wide variety of highly complex, but relatively cheap cameras and recorders. A potential drawback is that we are forced to follow the domestic market down whatever development route it is taking. This means using 625 (UK standard) or 525 (US standard) lines picture format and a corresponding loss in resolution compared to that of emulsion films. On a standard 625 line set, the image has a resolution less than 1/3 that of most emulsion films. This loss of resolution is compensated for by a high scanning rate which enables the detection of movement. The human eye is deceived by TV scanning owing to our ability to correlate from scan to scan subconsciously, identifying objects and rejecting much of the noise. Freeze-frame images (which would be used for detailed and quantitative analysis) reveal the flaws of TV and its relatively poor resolution, a feature that should not be overlooked. On the other hand, TV signals can be monitored in near real time; also, TV tubes can operate at extremely low light levels, and provide motion pictures.

Television (or photo-electric scanning) can be achieved using three principal types of receiver: Standard tube, Silicon Intensified Target (S.I.T.), and Charged Couple Device (C.C.D.). The standard TV consists of a target which is scanned by an electron beam. The target may be either photoconductive (e.g. the Vidicon) or photosensitive (e.g. the Image orthicon tube) and may detect images of 0.5 Lux (at the camera).

The S.I.T. tube consists of a silicon target vidicon-type tube incorporating an image-intensifier (Vigil, 1972). S.I.T. tubes can detect images down to 10^{-3} Lux. For even greater sensitivities an extra image-intensifier can be added to form an Intensified S.I.T. (I.S.I.T.) which can detect images down to 10^{-4} Lux (Harris, 1980).

The C.C.D. operates on a different principle, which does not use a scanning electron beam. Instead, the optical image is projected on to a target consisting of an array of photoelectric sensors. Scanning can be performed via a simple shift operation reading sequentially through the array, since the sensor elements are coupled electrically as a shift register. C.C.D.s have the advantage of wide dynamic range combined with rugged construction. The latest commercially available C.C.D. cameras can detect images of 2 Lux (Martin et al., 1987).

Direct comparisons between the effectiveness of TV tubes and films are not simple (see Patterson, 1981) as sensitivity, resolution and spectral response are interlinked and affected by factors such as exposure time and operating temperature. Figure 7 attempts to compare the sensitivity of various viewing systems in a semi-quantitative way.

When used with conventional (thallium iodide) light sources, and without filtering, TV cameras have similar range limitations to those of emulsion systems in the absence of LIBEC geometry (Hittleman et al., 1975). A possibility that has not been explored

yet is the combination of circular polarization and an S.I.T. camera. This combination may increase viewing ranges by a factor of two (Gilbert and Pernicka, 1966), bringing average operating altitudes up to 25 m for normal or in excess of 40 m for LIBEC geometry.

As components in a range-gated system, image intensifiers have the potential of significantly extending viewing ranges over conventional (emulsion) film techniques. The most important feature of an image intensifier in this respect is that it can be electronically gated with reasonably high precision. In the previous section the two range-gated systems under development were introduced. Only the Wilson (1986) system seems likely to become available, and despite its range limitation (40 m), in more turbid waters (on the continental shelf) it will enable photography in conditions that were previously unviewable.

The chief advantage of a TV system is that the data can be sent in near real time to the operator on board ship. In ocean floor-searching projects this is a prime consideration (Ryan and Rabushka, 1985), however it does bring its own problems. If data are to be sent direct to the ship, then a standard (or domestic) TV image requires more signal bandwidth than is available in a single element coaxial cable. Standard TV requires 4.5 MHz for transmission at a scan rate of 30 frames per second. A coaxial cable is capable of transmitting a bandwidth of 2.0 MHz over 8000 m (Mosley, 1988). Fibre optical towing cables have almost been developed to a point where they are available for deep ocean operations. For underwater TV photography the consequences are far reaching as these cables will eliminate all problems of data retrieval and storage. In the absence of fibre optic cables there are only two realistic options to solving the bandwidth problem. The first is to use a multicore cable. The second is to cut or compress the signal to fit the bandwidth available.

Signal cutting would involve slowing down the scanning rate or cutting down the number of lines. Neither option is attractive since reduced scanning would introduce flickering, and line reduction would cut resolution. In a system that is already at one third of the resolution of emulsion film and at long range from the sea floor, such measures could prove intolerable. In the case of range-gated systems, cutting may have already occurred (Wilson, 1986) as the laser cannot cope with a 50 Hz repetition rate. Most

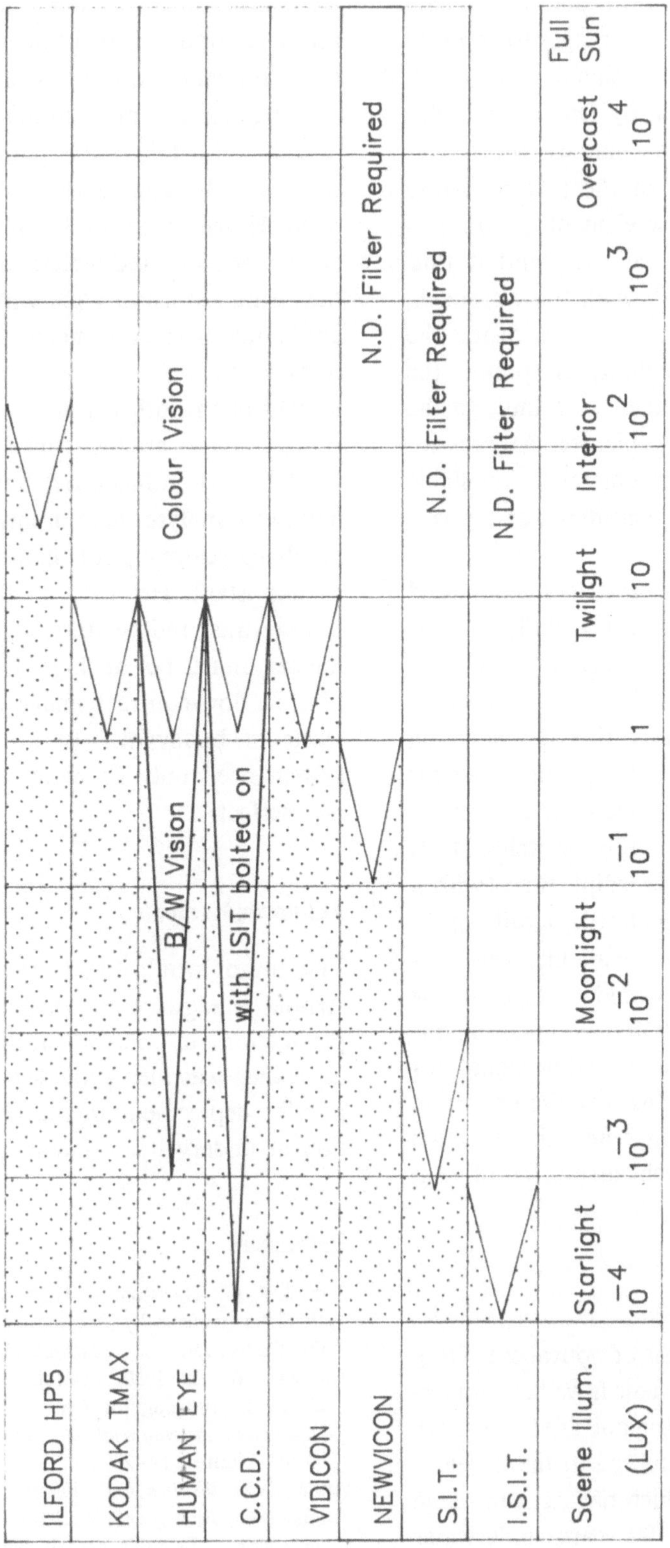

Fig. 7. Chart comparing viewing system sensitivities (after Harris 1980), see text for explanation.

underwater (emulsion) cameras have a finite recycling time of approximately 2 sec. If cutting the scan rate is the chosen option, then 2 sec could be adopted as the minimum scan time for TV images to compete with. A thorough discussion of signal cutting (i.e. narrowband television) is given by Deutsch (1968).

Signal compression is more complex but preferable. It involves the analysis of the picture before transmission so that only those elements of the picture that change between scans are transmitted, thus cutting the bandwidth required. With this approach, the more complex the target and the faster the system is towed over the sea floor, the poorer the pictures will become because there is a limit to the "compressibility" of a moving picture. At present, signal compression is not a commercially available option; however, it should be considered as a possible way forward.

Video recording using commercial systems placed in pressure housings has been successfully accomplished (Chezar and Lee, 1985). Even though the bandwidth of domestic VCRs is limited to between 2 and 3 MHz and the image resolution degraded on recording (by approximately 50%), video systems are cheap and reliable. A video recorder can record up to 4 hours of signal or can be switched to a half-cut signal to double its recording time (with a corresponding loss of information). Until signal compression techniques become available, one solution to TV viewing problems would be to store the full signal on one or more VCRs mounted on the towed vehicle and to cut the monitor signal for transmission up to the ship. This way the operation of the vehicle can be monitored and the eventual replay (after retrieval of the vehicle) can be of relatively high quality.

Summary

Since the late 1960s the problems of optical scattering and signal transmission in seawater have been understood but until recently no original advances have been made in either field. Compared to the development of aerial photography which has risen in range from 1000 to 100 000 m over 100 years, underwater photography has struggled through its physical limitations to an increase in range from 4 to 40 m.

Cocking (1976) pointed out that images viewed on standard TV screens are between $\frac{1}{6}$ and $\frac{1}{3}$ the resolu-

tion of emulsion film. The problems of signal storage and transmission combined with the relatively poor resolution of TV systems make emulsion films an attractive option if real-time monitoring and extreme (>21 m) ranges are not a prime requirement.

Range-gating may ultimately become the best technique for long-range photography, as higher resolution and higher sensitivity TV systems are developed. However, at present photography in relatively clear, deep oceanic waters is probably best carried out using emulsion films with polarizing filters and arc lamps. If video cameras are to be used for image retrieval then the problems of poorer resolution and limited bandwidth will have to be accepted. At long ranges, images are low-contrast and any reduction in resolution or transmitted bandwidth will affect the ability to analyse them in any detail. For operations involving searching for man-made objects this is not so important. However, in the study of subtle and low-contrast sedimentary features on the sea floor, images must be of high quality. Transmitting a blurred, low-contrast image at high resolution is a waste of bandwidth. In practice, the transmission system need only be better than the source image, not perfect.

Acknowledgements

The author would like to thank Mike Conquer and Arnold Madgwick for their advice and help in the production of underwater photographs. Invaluable assistance was also provided by the IOSDL applied physics department in the construction and operation of underwater cameras.

References

Aldred, R. G., Thurston, M. H., Rice, A. L., and Morley, D. R., 1976, An Acoustically Monitored Opening and Closing Epibenthic Sledge, *Deep-Sea Research* **23**, 167–174.

Andrews, W. B., 1980, Inspection, Maintenance and Repair: Ultrasonic Imaging for Offshore Underwater Operations, in *Oceanology International '80*, Technical Session J; B.P.S. Exhibitions Limited, 19–28.

Anon., 1953, *Wratten Light Filters*, Kodak, London, 92.

Anon., 1968, *Bibliography on Underwater Photography and Photogrammetry*, Eastman Kodak, New York, 23.

Anon., 1986, Underwater Cameras-Product Review, *International Underwater Systems Design* **8**(5), 6–8.

Anon., 1987, Current Trends in Underwater Photography; A Company Perspective, *International Underwater Systems Design* **9**(1), 30 pp.

Ballard, R. D. and Moore, J. G., 1977, *Photographic Atlas of the Mid-Atlantic Ridge Rift Valley*, Springer-Verlag, New York, 114 pp.

Briggs, R. O. and Hatchett, G. L., 1965, Techniques for Improving Underwater Visibility with Video Equipment, *Ocean Science and Ocean Engineering* **2**, 1284–1308.

Brown, J. F., 1985, The First Underwater Photograph, *The British Journal of Photography* **32**, 894–895.

Chardy, P., Guennegan, Y., and Brannelec, J., 1980, Photographie sous-marine et analyse des peuplements benthiques. *C.N.E.X.O.*, *Rapports Scientifiques et Techniques* **41**, 32 pp.

Chezar, H. and Lee, J., 1985, A New Look at Deep Sea Video, *Deep-Sea Research* **32**(11A), 1429–1436.

Cocking, S. J., 1976, Improving Underwater Viewing, (139–190) in E. A. Drew, J. N. Lythgoe and J. D. Woods, (eds.), *Underwater Research*, Academic Press, London. 430 pp.

Deutsch, S., 1986, Narrowband Television Pictures for Transmission via Oceanographic Sound Waves, *Ocean Engineering* **1**, 9–16.

Dixon, T. H., Pivirotto, T. J., Chapman, R. F., and Tyce, R. C., 1984, A Range-Gated Laser System for Ocean Floor Imaging, *Marine Tech. Soc. Journal* **17**(4), 5–12.

Duntley, S. Q., 1963, Light in the Sea, *Journal of the Optical Soc. of America* **53**(2) 214–233.

Fritz, N. L., Specht, M. R., and Needler, D. G., 1972, A New Colour Film for Water-Penetration Photography, Preprints of the *8th Annual Conference of the Marine Technology Society 1972*, pp. 737–738.

Funk, C. J., 1973, Predicted System Performance of Improved Underwater Light Sources. (pp. 7–17) in Anon (ed.), *OCEAN '73 Engineering in the Ocean Environment*, IEEE, New York, 623 pp.

Gilbert, G. D. and Pernicka, J. C., 1966, Improvement of Underwater Visibility by Reduction of Backscatter with a Circular Polarization Technique, in *Underwater Photo-Optics: Seminar Proceedings*, Society for Photo-Optical Instrumentation Engineers, Redondo Beach, Ca. pp. A-III-1 to 11.

Harford, J., 1968, Underwater Lighting—A Status Report, in F. Alt, (ed.), *Marine Sciences Instrumentation*, *Vol. 4*. Plenum, New York, pp. 373–380.

Harris, R. J. 1980, Improving the Design of Underwater TV Cameras, *Underwater Systems Design* **2**(1), 7–11.

Harris, S. E. and Ballard, R. D., 1987, ARGO: Capabilities for Deep Ocean Exploration, *OCEANS '86 Conference Record*, Vol. 1 MTS, Washington, pp. 6–8.

Heckman, P. J., Jr., 1966, Underwater Range-gated Photography, pp. B-IX-1 to 9 in *Underwater Photo-Optics*: Seminar Proceedings; Society of Photo-optical Instrumentation Engineers. Redondo Beach, Ca.

Heezen, B. C. and Hollister, C. D., 1971, *The Face of the Deep*, Oxford University Press, 659 pp.

Hittleman, R. L. and Strickland, C. L., 1968, The Application of Incandescent, Mercury Vapour, and Thallium Oxide Lighting to Underwater Tasks, Society of Photo-optical Instrumentation Engineers. *Seminar Proceedings Vol. 12*, pp. 153–165.

Hittleman, R. L., Vigil, A. E., Coffman, J. E., and Hatchett, G. L., 1975, The Performance of Low-Light Level Television Cameras on Underwater Remote Controlled Vehicles and Towed Sensor Platforms, *S.P.I.E. Vol. 64*, pp. 128–134.

Huggett, Q. J., 1987, Mapping of Hemipelagic versus Turbiditic Muds by Feeding Traces Observed in Deep-Sea Photographs, in P. P. E. Weaver and J. Thomson, (eds.), 1987, *Geology and Geochemistry of Abyssal Plains*, Geol. Soc. Spec. Pub. No. 31, pp. 105–112.

Jones, C. H. and Gilmour, G. A., 1976, Sonic Cameras, *J. Acoust. Soc. Am.* **59**(1), 74–85.

Lampitt, R. S. and Burnham, M. P., 1983, A Free Fall Time-Lapse Camera and Current Meter System 'BATHYSNAP' with Notes on the Foraging Behavior of a Bathyal Decapod Shrimp, *Deep Sea Research* **30**, 1009–1017.

Laughton, A. S., 1957, A New Deep-Sea Camera, *Deep-Sea Research* **4**, 120–125.

Martin, G. J., Womack, K. H., and Fischer, J. H., 1987, A High Resolution CCD Camera for Scientific and Industrial Imaging Applications, *Proc. Society of Photo-optical Instrumentation Engineers* **818**, 301–319.

Mauviel, A., 1982, La bioturbation actuelle dans le milieu abyssal de l'Océan Atlantique Nord, Thèse, Diplôme de Docteur de troisième cycle, Université de Bretagne Occidentale, 103 pp.

Mertens, L. M., 1970, *In-Water Photography: Theory and Practice*, John Wiley & Sons, New York, 391 pp.

Mosley, C., 1988, Video Systems in Deep Water Mining and Oceanographic Applications, Underwater Systems Design **10**(4), 20–22.

Nathan, A. M., 1957, *A Polarizing Technique for Seeing through Fogs with Active Optical Systems*, New York University, Tech. Rep. 362.01.

Ohta, S., 1983, Photographic Census of Large Sized Benthic Organisms in the Bathyal Zone of Suruga Bay, Central Japan, *Bull. Ocean Res. Inst.*, Univ. Tokyo, No. 15, 244 pp.

Patterson, E., 1981, Underwater Television—An Art in a State, *International Underwater Systems Design* **3**(5), 11–12.

Patterson, R. B., 1972, Increased Ranges for Conventional Underwater Cameras, *Proc. S.P.I.E.* **24**, 153–161.

Phillips, J. D., Driscoll, A. H., Peal, K. R., Marquet, W. M., and Owen, D. M., 1979, A New Undersea Geological Survey Tool: ANGUS, *Deep-Sea Research* **26**(A), 211–226.

Rebikoff, D., 1967, History of Underwater Photography, *Photogrammetric Engineering* **33**(8), 897–904.

Rice, A. L. and Collins, E. P., 1985, The Use of Photography in Deep-Sea Benthic Biology at the Institute of Oceanographic Sciences, (pp. 153–164) in J. D. George, G. I. Lythgoe and J. N. Lythgoe (eds.), *Underwater Photography and Television for Scientists*, Clarendon, Oxford, 184 pp.

Rice, A. L., Aldred, R. G., Billett, D. S. M., and Thurston, M. H., 1979, *The Combined Use of an Epibenthic Sledge and a Deep-Sea Camera to give Quantitative Relevance to Macro-Benthos Samples*. AMBIO Special Report No. 6, pp. 59–72.

Ryan, P. R. and Rabushka, A., 1985, The Discovery of the Titanic by the U.S. and French Expedition, *Oceanus* **28**(4), 16–33.

Vigil, A. E., 1972, A New Low Light Level Television Camera for Underwater Applications, Preprints of the *8th Annual Conference of the Marine Technology Society, 1972*, pp. 581–600.

Wal, M. R., 1968, Underwater Viewing, *Proceedings Society of Photo-Optical Instrumentation Engineers* **12**, pp. 21–32.

Wilson, T. B., 1986, An Experimental Laser Gated-Television Underwater Viewing System, *Proceedings of Electro-Optics and Laser, U.K. '86*, Published by Communications in Print, Essex UK.

PART II

Sampling Techniques

Current Methods for Obtaining, Logging and Splitting Marine Sediment Cores

P. P. E. WEAVER and P. J. SCHULTHEISS *

Insititue of Oceanographic Sciences (DL) Brook Road, Wormley, Godalming, Surrey GU8 5UB, UK

(Received 27 April, 1989; accepted 1 September, 1989)

Key words: coring devices, box corer, gravity corer, piston corer, giant piston corer, 'P' wave log, wholecore logging, kevlar, warp.

Abstract. The main types of deep-sea sediment coring devices are described and their relative merits and drawbacks are discussed. These devices include box corers, gravity corers, piston corers, giant piston corers and vibrocorers. Recent utilisations of kevlar and polyester coring warps are also discussed, since these are the only warps capable of handling the large weights associated with the larger devices. Recent developments in wholecore logging, including 'P' wave, density and magnetic susceptibility, are described as are methods of subcoring and core splitting to obtain the maximum amount of detail on the split surfaces. The wholecore logs together with a good colour photograph of the recently split sediment surface provide a lasting unambiguous record of the core.

Introduction

Our knowledge of the deep-sea floor is built up from interpretations of a combination of imaging techniques (geophysical mapping and profiling) and by direct sampling. A vast range of devices has been developed for sampling the seabed, and these fall into various categories, with each category being aimed at either particular depth penetrations, particular sediment types or a specific use for the sample, The deepest and most complete record of deep-sea sediments and basement rocks has been provided by the Ocean Drilling Program (ODP), using a combination of downhole hydraulic piston coring and rotary drilling. The drilling capabilities of the ODP and its predecessor the Deep Sea Drilling Project (DSDP) have been well documented elsewhere (see, Storms, this volume) and are not considered further in this chapter. Smaller and less expensive research vessels than those used for drilling are capable of

---

* Present address: Schultheiss Geotek, Fern Cottage, Marley Lane, Haslemere Surrey GU27 3RF, UK

using a variety of devices to sample the upper 20–25 m of the sediment column. The largest of these ships are also capable of handling the relatively new giant piston corers which currently can sample to depths of 35 m below the sea floor, and in the future may be able to sample as deep as 50 m.

There are two important considerations to be taken into account before taking a deep-sea core: (a) how hard is the substrate?; and (b) what is the core to be used for? In some circumstances it may be prudent or necessary to use more than one type of corer to ensure that a complete record of a site has been obtained to the depth of interest or to a depth limited by the available technology. This is often the case if a good-quality core of the sediment surface is required together with a sediment record several metres long. Scientists wishing to make physical property measurements, and those interested in an accurate record of the thicknesses of all lithological units in a core, should be particularly careful and aware of problems arising from core disturbance and core mis-sampling. In some cases these are not easily detected and may lead to incorrect interpretations.

Coring techniques have developed over the years with a lot of ingenuity but without any significant "high technology". The success of a coring operation has all too often been judged by the criteria of "how full or how long is it?". Attention has, more recently, been focused on the problems associated with core disturbance and ensuring the general quality and representativeness of the core rather than solely that of maximizing the depth of penetration. There has been a move towards the design of more sophisticated corers in recent years, often involving the mounting of monitoring equipment on the corer, both to improve penetration and to produce a detailed record of that penetration. Although some work has been completed on the design and testing of this type of equipment (see Parker and Sills, this volume), there is no operational corer which uses it routinely.

Several other major reviews of coring equipment (e.g. those of Bouma, 1969; Moore and Heath, 1978; Lee and Clausner, 1979) and of core handling (Bouma, 1969) have been published in the past two decades. Details of several corers not mentioned in this chapter will be found in those papers. This chapter summarizes the main types of coring equipment currently in use, including recently developed giant piston corers, as well as modern methods for handling and describing cores. There is now a greater emphasis on whole core logging and high-quality core photography, which can replace and improve much of the laborious core description work frequently carried out by inexperienced personnel.

Box Corers

The main advantage of box corers is that they obtain large-volume cores of surface sediment with minimum disturbance. They consist of a square box (occasionally a large diameter cylinder), a head-weight and a spade-type lever arm (Fig. 1a). Some designs also include a tripod support frame to ensure vertical coring (Fig. 1b). The cross-sectional area of the box may be up to $0.25\,\mathrm{m^2}$, but the length is generally less than 1.2 m. The spade lever arm lies horizontally during deployment and descent, but on recovery it is pulled into its vertical position, thus closing the bottom of the core box prior to pull-out. In many of the designs the spade lever arm also closes the top of the box, thus ensuring an undisturbed sediment–water interface. Some box corers utilize a scissor arrangement of two spade lever arms (Fig. 1a) which ensures that the box is maintained in a vertical position, thus eliminating the need for a tripod frame. It has become apparent, however, that single-spade corers are often much better at coring sandy sediments. Most box corers activate the spade closure by means of a "no-load" release. Although this system is simple, it can lead to pre-tripping, especially in rough seas. The IOS box corer (Fig. 1a) has an acoustically activated release which is fired after the corer has

penetrated the sediment. Although this system potentially allows the corer to bounce on the sea floor, no problems have been experienced to date.

Gravity Corers

These corers are very simple, but variable in design, consisting of a large headweight, to which is attached a barrel of variable length with a core cutter and catcher at the lower end. The ship's warp is attached directly to the core head, except in the device described by Hvorslev and Stetson (1946), in which the corer is triggered by a trigger weight. Most round barrel corers of this type (Fig. 1c) utilize core liners, but those with long box-shaped barrels, such as the Kastenlot corer (Fig. 1d), do not. Most incorporate a flap valve at the top to allow water to escape during coring but which is closed during pullout, ascent and recovery to prevent sediment being washed out. Problems with washout occur particularly at the sea surface during retrieval as a result of the loss of buoyancy when the core is lifted from the water.

Gravity corers are inexpensive and easy to use and can take high-quality cores of the upper few metres of the sea floor. They are, however, subject to two major sources of error. These are (1) mis-sampling due to sediment plugging in the core barrel, and (2) repenetration caused by vertical oscillations of the ship's warp. Emery and Dietz (1941) showed that open barrel gravity corers were capable of taking very shortened sections of core, and Hvorslev and Stetson (1946) showed how the sediment layers would be shortened ahead of the corer. Weaver and Schultheiss (1983b) were able to show evidence of gravity corers "bouncing" on the sea floor, enabling multiple sampling of the upper sediment section by repeated penetrations. They also showed how each successive repenetration was shortened relative to the previous one, and how the shortening was more pronounced in the softer more plastic, sediment layers. This suggests that gravity corers can be unreliable in sediments of mixed composition, and produce increasingly unreliable results with increasing penetration. The length of core that can be taken with an open barrel gravity corer is limited by the increasing friction between the cored sediment and the inside of the barrel. At some stage the force required to move the core up the barrel

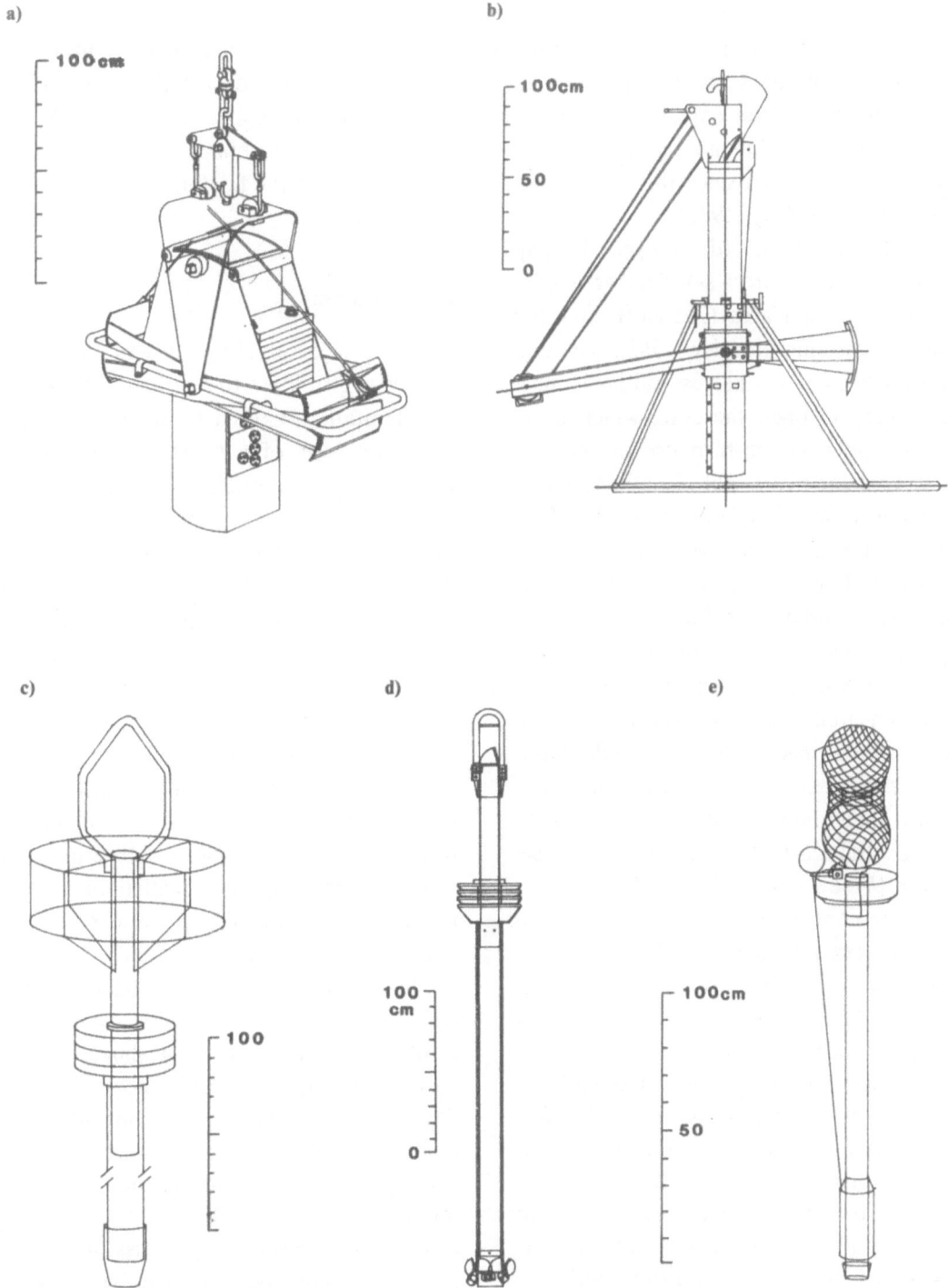

Fig. 1. Corers used for near-surface sediment sampling: (a) IOS box corer (double spade) (after Peters *et al.*, 1980); (b) box core with single-spade and tripod frame (see Bouma, 1969); (c) hydrostatic gravity corer (Richards and Keller, 1961); (d) square barrel Kastenlot core (Kogler, 1963); (e) free-fall, pop-up boomerang corer (Moore, 1961).

exceeds the force required for the barrel and core to act as a solid rod for further penetration. This is known as "plugging" and commonly occurs between 3 and 6 m penetration. Intermittent or partial plugging is the cause of core shortening. A more detailed study of this effect has been made by Parker and Sills (this volume) in one sediment type using a technique to continuously monitor the sediment surface both inside and outside the core barrel during penetration.

Gravity cores are frequently used as trigger weights for piston corers and often produce a more representative record of the surface sediment than that obtained from the piston core. Care must be taken, however, to ensure that no repenetration of the trigger corer has occurred since this is common during piston coring (McCoy, 1980).

The Kastenlot corer (Fig. 1d) bridges the gap between gravity and box corers since it has a relatively large cross-sectional area (225 cm²), and can take cores up to 6 m long (Kogler, 1963). The box splits longitudinally into two halves and so it must be laid horizontally to open the core, which often causes some slumping and disturbance of the surface sediment. This corer does not use a liner and is usually sub-sampled immediately after opening. Sections of the cored material can be stored in 1 m-long boxes, which have the same cross-section as the corer but with closed ends for later examination. It is also possible to remove one corner of the core into a V-shaped trough which can be stored in a "D" tube as an archive section. The large cross-sectional area allows this corer to provide high-quality cores which often suffer less disturbance and core shortening than those taken by other gravity corers. However, the Kastenlot corer is susceptible to repenetration (Weaver and Schultheiss, 1983b).

The Boomerang corer (Fig. 1e) is a modification of the open-barrel gravity corer which operates as a freefall/pop-up corer. It consists of a ballast section comprising a steel barrel, weight, steel float protection and lead pilot weight. The float section consists of a core liner, valve release and two glass spheres. The whole system is released from the ship, after which it quickly rights itself and attains a terminal velocity of approximately 7 m/sec. A hollow rubber ball prevents release of the float portion during deployment, but this is compressed and released during descent. When the corer impacts with the seabed the pilot weight slides up the barrel, releasing the float section which then ascends, closing the valve-release at the top of the liner and pulling free the liner tube with the core inside. The float section ascends through the water column and is identified at the surface by an electronic flash in one of the spheres. The advantage of this system is that several cores can be obtained rapidly by deploying them as a series prior to recovery. Furthermore, the core site is likely to be directly beneath the ships position,

since the round-trip time is about 15 min/1000 m water depth and hence there is little time for the equipment to drift during the operation. The disadvantage of the system is that only short cores (<1.2 m) can be obtained and it can sometimes be difficult to find the floating core in poor visibility or rough weather.

Piston Corers

The piston corer (Fig. 2) was first developed by Kullenberg in 1947 to overcome the depth penetration limitation, caused by plugging, of gravity corers. It consists of a series of connected barrels, a large, heavy headweight, a trigger arm and a trigger corer. A plastic core liner is usually employed, but the core may be extruded on deck after recovery. The head may be designed with removeable weights, with fins, or with housings for instruments such as cameras, flash-guns or pingers. Most trigger arms are fitted with safety pins to prevent accidents during deployment. Hydrostatic pins usually retract in the upper or mid water column, but acoustic releases can be used, which can prime the corer by command from the ship at any depth. These offer increased safety in rough weather or on less stable ships, where temporary reductions in load on the trigger arm due to ships' heave can cause premature release of the corer.

The advantages of the piston corer over gravity corers is the increased depth of penetration as a result of the prevention of plugging. This should ensure higher quality cores. The depth of penetration depends on sediment type, but in relatively soft muds, over 20 m can be obtained (Kuijpers et al., 1984). The action of the piston, which reduces the internal friction, enables these corers to generally recover more complete, and less disturbed sediment sequences than open-barrel gravity corers. The attachment of the main warp to the piston, however, causes ships' heave motions and elastic rebound of the warp after corer release to be transmitted to the piston during coring, which can result in core shortening and/or other sediment disturbances (McCoy, 1980, 1985). Deformation is often particularly severe in the vicinity of sandy layers which may be tilted, bent downwards at the sides of the liner or totally mixed by the coring action. Igarashi et al. (1970) showed that coarse particles could be dragged from

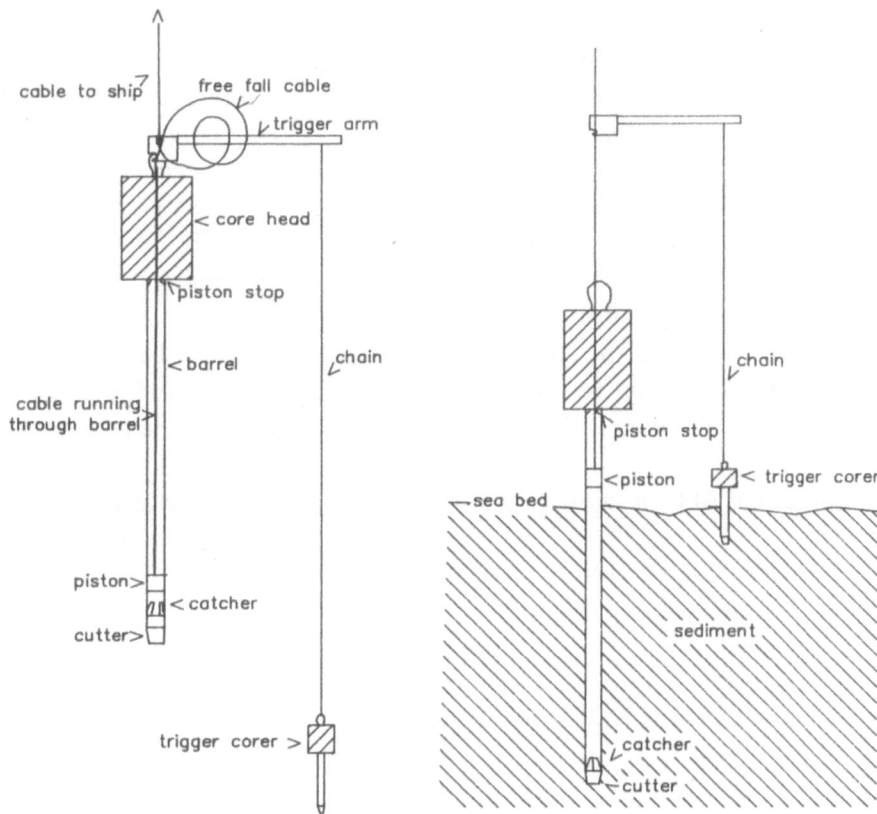

Fig. 2. Typical piston corer shown (a) as deployed and (b) after triggering at the sea-floor.

the sediment surface and embedded in the deeper parts of the core without there being obvious visual disturbance to the core. Burns (1962) showed that the upper sediment layers could be mis-sampled by improper adjustments of the trip chain length or incomplete immobilization of the piston during coring.

The sediment surface is often lost in piston cores, and if the relative lengths of freefall and trip-chain are not calculated correctly, the upper few metres of the sediment column can be omitted. Trigger cores usually provide a more accurate record of the surface sediment, but care must be taken in their interpretation because of the repenetration problems mentioned previously. The piston may be of a solid or breakaway type; the breakaway version splits into two at the beginning of the pull-out and thus prevents material flowing into the core barrel if it is not completely full.

The majority of piston corers have internal liner diameters of 65–70 mm with many being 66 mm, the same as used by the ODP. The Geological Survey of the Netherlands, however, have experimented with liner diameters of 83 mm and 145 mm, and have found that 83 mm liners give good-quality cores of comparable length to the small-diameter liners, but obviously with much larger volumes of material retrieved, (Kuijpers et al., 1984). This can be particularly useful when sediments are required for geotechnical and geochemical studies.

Giant Piston Corers

THE ADVANCED PISTON CORER (APC)

The concept of a long piston corer, capable of penetrating up to 50 m of sediment was first proposed by the Wood's Hole Oceanographic Institution in 1967, and reports of the first trials were published by Hollister et al. (1973) and Driscoll and Hollister (1974). The first model was called the "Giant Piston Corer" (GPC) and took sediment samples 110 mm in diameter. An account of the design and early modifications of this corer is given by Hollister et al. (1973). The most serious problem was encountered with the core barrels, which were standard well casing pipes that frequently bent or broke at the barrel couplings on 30–35 m deployments. Further problems were encountered with stabilizing the piston and tripping the core. Driscoll and Hollister (1974) discussed several modifications to the system, including step tapered barrels, a redesigned head with provision for instrumentation and an acoustic release. In 1977 a major redesign was proposed (Driscoll and Silva, 1977) to incorporate these features into a technically sophisticated system for taking 50 m-long cores. Initially called the Long Coring Facility, it was later renamed the Advanced Piston Corer (APC). (This should not be confused with the Ocean Drilling Program's hydraulic "APC".)

The instrumentation of the APC head includes an acoustic release, a tilt alarm, a rotation recorder, a piston control system, an acoustic telemetry package and a power and control package. These instruments were designed to enable close control and monitoring of the system during the coring operation. A simpler version of this system (Fig. 3) was used on the Giant Piston Corer (Silva *et al.*, 1977) and on the *Marion Dufresne* ESOPE expedition (Schuttenhelm, in press). In the simple version, a parachute is attached to the coring warp just above the core head, the warp continues down the barrels to a standard-type piston and is held in position at the core head by the release. The parachute takes some of the core weight during lowering, but on approach to the seabed the winch payout is speeded up so that the parachute

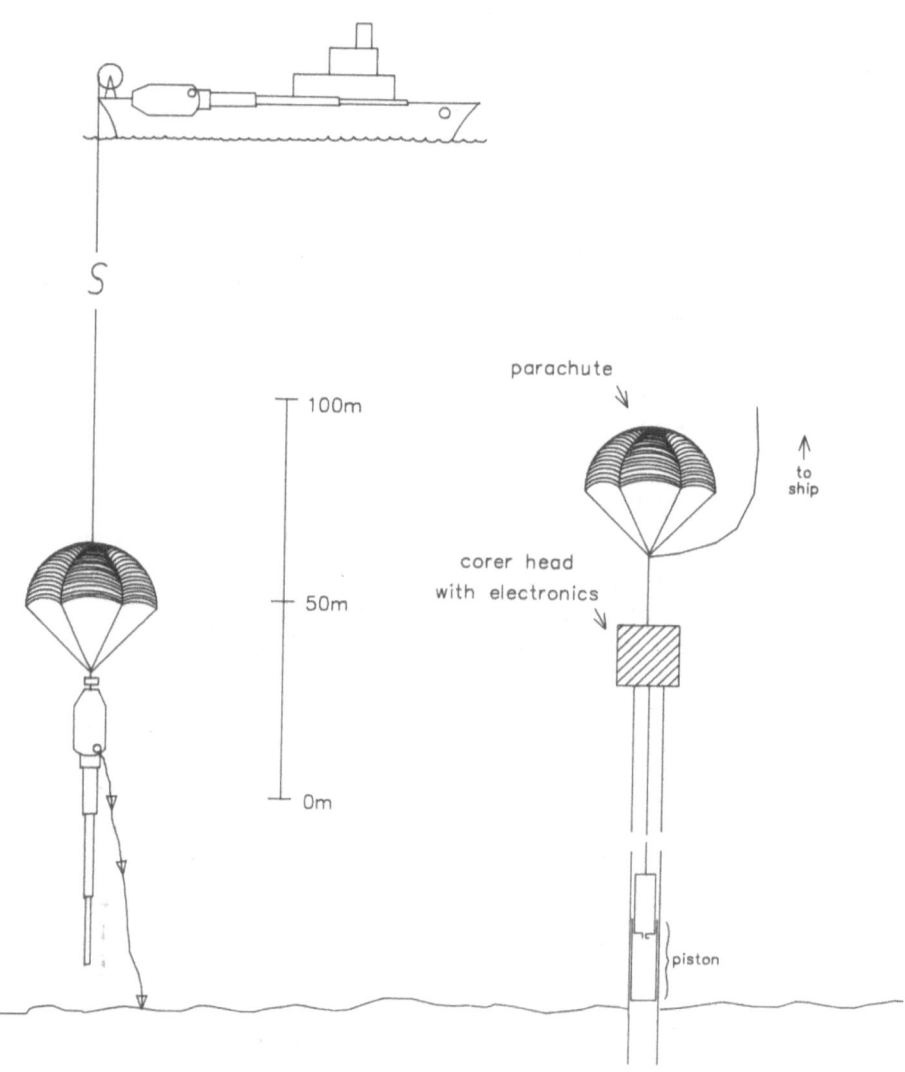

Fig. 3. The giant piston corer with parachute (a) during descent; (b) during penetration (redrawn after Driscoll, 1981).

takes the whole weight of the corer, thus producing a quantity of loose warp and eliminating any rebound effects of the warp when the load is released. The core is tripped with minimal freefall, and slides past the piston which is held at the same position by the slowly descending parachute.

A very complex system has been considered for stabilizing the piston by decoupling it from the main ship's warp (Driscoll, 1981). Although this system has never been built, it illustrates the extent to which technology can be utilized in coring equipment. It is known as the Hydrostatic Accumulator Piston (HAP), and utilizes a parachute located just above the core head. The HAP piston concept consists of two long chambers, the upper one air-filled and connected to the lower oil-filled one by a valve. There is also an accelerometer to assess the piston motion and a sonar transducer at the lower end to monitor the position of the sediment surface during coring. The piston is connected to the parachute as before, but after triggering the valve between the oil and air filled chambers allows oil to pass, thus compressing the length of the piston. The rate of contraction of the piston is designed to keep the base of the piston stable during penetration (which takes only a few seconds). The contraction rate needed can be calculated from the parachute sink rate, or from the accelerometer which monitors the vertical motion of the bottom part of the piston.

The drawbacks of the APC system are its great complexity and the very limited numbers of ships capable of dealing with the large loads experienced during pull-out.

STACOR

The only other corer which has taken cores in excess of 25 m is the STACOR (stationary piston corer), developed by the Institut Français du Pétrole, Elf Aquitaine and Total in the mid-1970s (Montarges *et al.*, 1983, 1987). This corer utilizes a mechanical system for maintaining a stationary piston, an idea first suggested by Kullenberg (1955) and later developed by Kermabon and Cortis (1968). The principle of the stationary or recoilless piston is that it is coupled to a baseplate on the outside of the corer via a series of pulleys and wires (Fig. 4). The main ship's warp connects directly with the corer head and is thus decoupled from the piston. The baseplate and piston begin the coring operation at the cutter end of

the barrel, and since the baseplate is 1.5 m across, it remains at the sediment surface and holds the piston at the same position while the corer slides past. This corer takes cores with a 110 mm internal diameter and, because of the stationary piston, generally achieves recovery approaching 100% with excellent core quality.

The STACOR system does not depend on sophisticated technology and has a high success rate. It has so far been launched in three different ways: over the stern of a ship, over the side rail and via a moon pool (Montarges *et al.*, 1987). The long barrel-length and the large loads, particularly those associated with pull-out (about 200 kN) restrict its use to the larger oceanographic vessels, or to vessels such as the *Nadir*, which have a large free deck for deployment over the stern. The loads can be drastically reduced by using Kevlar cables which are almost weightless in water (Schilling *et al.*, 1988). One other drawback with the current STACOR system is the length of time required to complete a coring operation, which can be as much as 16 hours. A large proportion of this time is spent removing the core from the barrel, which detracts from the corer's usefulness for geochemical work where cores often need to be placed rapidly under nitrogen (Schuttenhelm *et al.*, in press). However, some redesign features of the system could probably overcome these drawbacks.

Vibrocoring and Rockdrilling

The vibrocorer extends the range of sediments which can be sampled to include stiff and stony clays, soft rock and sands, all of which are difficult to penetrate using conventional gravity or piston corers. The British Geological Survey's vibrocorer (Fig. 5) consists of a tripod frame with a base-mounted rotary drive table which drives a 6 m-long hexagonal drill barrel. The twin vibrator motor gives a force of 6 tonnes at 50 Hz and the core can be withdrawn from the sediment by a base-mounted winch capable of exerting 12 tonnes force. This corer takes 83 mm diameter cores in plastic liners. The system is electrically powered from the ship, from where the operation can be monitored and controlled (Ardus *et al.*, 1982). The present system has been tested to 1800 m water depth, with the main limiting factor being the power supply cable.

The British Geological Survey's rockdrill is

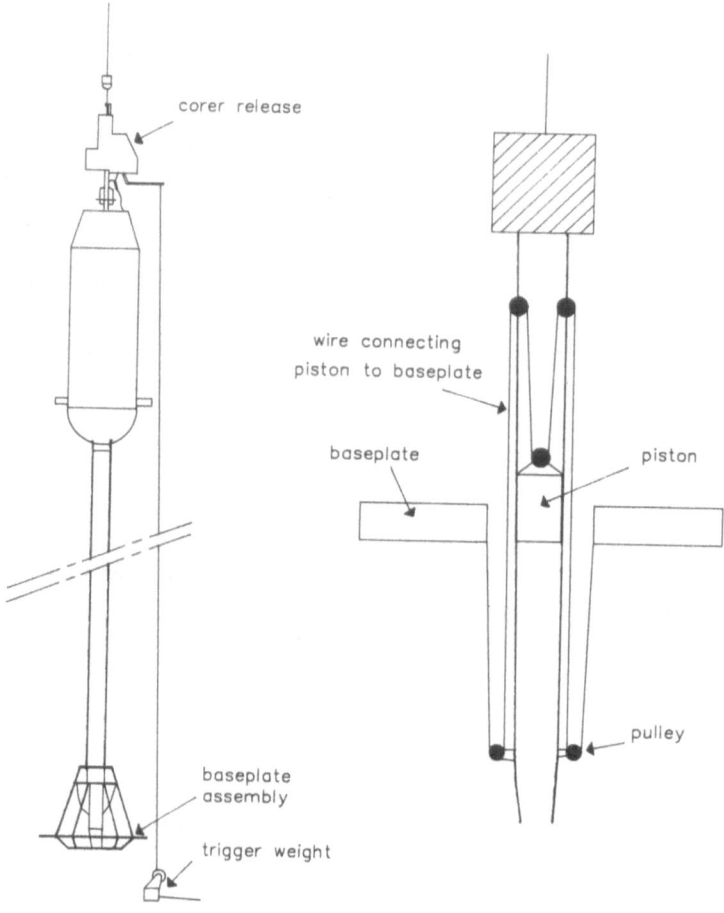

Fig. 4. The stationary piston corer, Stacor, showing the operating principle of how the base-plate and piston remain at the sea-floor during penetration (redrawn after Montarges *et al.*, 1983).

mounted in the same tripod structure as the vibro-corer (Pheasant, 1984). Up to 6 m penetration is possible with this system and, as before, withdrawal of the core is achieved by the 12-tonne winch. Brook and Pelletier (1970) report a deep-sea rock drill which works by hydraulic power, but which is also limited to 1800 m water depth. This system utilizes three heavy-duty gas cylinders which allow water to enter by hydrostatic pressure which turns a hydrostatic motor for a 7.5-minute cycle. The first half of the cycle drills the core in and flushes the drill bit and the second half of the cycle withdraws the core.

Winches and Warps

The maximum loads experienced during coring operations are those associated with pulling the corer out of the sediment. For most gravity, box and piston coring work the loads will be generally within the capability of most oceanographic winches. The desire to produce piston corers of larger diameter, and giant piston corers has, however, increased the potential maximum loads to such an extent that conventional steel warps are no longer adequate. The problem with steel warps is their large weight in water, so that in deep water, beyond a certain limit the increased strength of a warp is largely used to offset the increased weight of the warp itself.

Two new materials have been used in recent years to overcome this problem; Kevlar and braided polyester. Both of these materials have negligible weight in water. The Geological Survey of the Netherlands has experimented with Kevlar and found that it gives excellent results when used with piston corers of large and small diameter. The Kevlar is encased in a polyethylene sheath to prevent chaffing and is stored on a drum like steel warp (Schilling *et al.*, 1988). The polyester cable was

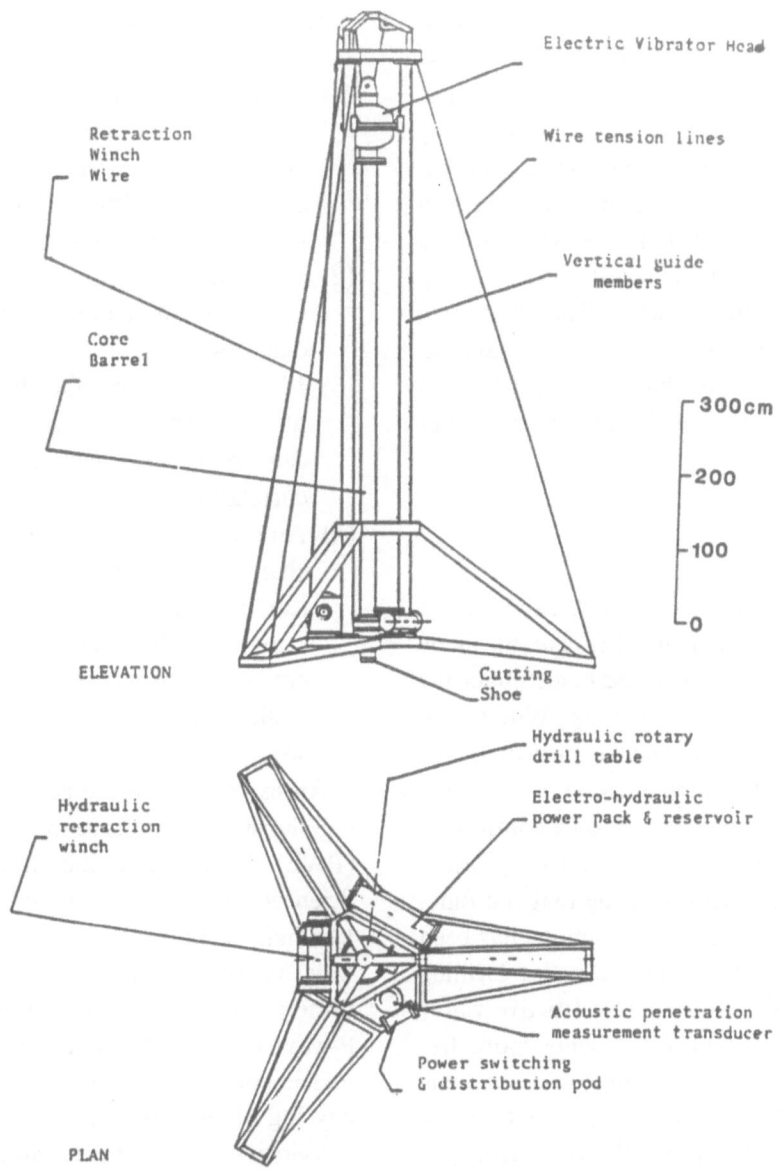

Fig. 5. The British Geological Survey's vibrocorer/rotary drill (from Pheasant, 1984).

chosen for the APC corer (Dzwilewski and Driscoll, 1980), even though this cable cannot be stored on a drum because it requires a long time for strain recovery after it has been loaded. This strain recovery would exert tremendous force on the drum and so the cable is stored loose in a container on deck (Driscoll, 1981). Both types of cable stretch significantly under load, which must be allowed for when calculating the length of wire required for the free-fall distances of piston corers. Both of these cables have considerable advantages over steel cables: they are light in water; they have smaller bending radiuses

than steel; they do not corrode; they can be cut and spliced, and yet Kevlar is stronger than a steel cable of equivalent diameter and polyester has about half the strength of steel. Polyester has about the same cost as steel for equivalent strength cable, but it has the disadvantage of needing a long recovery time after it has been stressed, and it cannot, therefore be wound on a drum. Kevlar, by contrast, costs about three times as much as steel, but it can be stored on a drum.

Many oceanographic cruises today use charter ships which may not have the necessary combination of winch, warp and "A" frame required for piston coring. Houbolt (1971) described a system combining all these features which could be transferred from ship to ship. The Houbolt winch has been considerably improved and is now marketed by the Dutch company Seabed BV. The winch is completely independent of the ship, having its own diesel hydraulic power pack, 10 000 nm of Kevlar cable, a telescopic boom, a traction unit with 20-ton lifting capacity, and two handling winches. The whole system has the dimensions and fittings of a standard container and can thus be freighted around the world and bolted on to any ship with suitable container anchorage points.

Whole Core Logging

Having obtained a core in a plastic liner, there are a number of useful measurements that can be made prior to sectioning or splitting. These come under the general heading of whole-core logging. Whole-core logging is defined as non-destructive measurements that can be automatically made on cores within plastic core liners at frequent intervals or on a continuous basis. The most frequently used is gamma ray attenuation which essentially provides a log that is a function of the sediment density. This technique has been used on cores obtained from the Deep Sea Drilling Project for many years. Other non-destructive whole core logging techniques which are being more frequently used are P-wave velocity logs and magnetic susceptibility. It can be argued that much data is lost from many sediment cores (especially those from the ocean drilling legs) because the time constraints onboard ship preclude many measurements being made at appropriately small sampling intervals. Perhaps any useful parameter that can be measured continuously and rapidly and does not require extra personnel should be measured on a routine basis.

The potential value of some whole-core logs is already clear (Schultheiss and Mienert, 1988); other non-destructive logging techniques may produce data of only limited intrinsic value, or of a value that is currently unclear. However, while any single logged parameter may at worst be of marginal value, the combination of logs is almost certain to provide a valuable diagnostic data set in the same way that down-hole logs are most useful when used as combination logs.

Continuous whole-core logging, with a range of different sensors would be of considerable scientific value for many reasons:

(1) Complete continuous records of P-wave velocity and density (from gamma ray attenuation) are necessary to construct synthetic seismograms for comparison with seismic reflection profiling records and borehole velocity logs.

(2) Continuous density logs are needed to evaluate the state of compaction of sedimentary sequences.

(3) Magnetic susceptibility logs provide rapid identification of terrestrially derived sediments and can provide excellent, high resolution, interhole correlations.

(4) Any logged parameter that is sensitive to subtle changes in sediment type or sediment structure can be invaluable as a guide for later detailed sampling. They may, for example, reveal cyclicity which is caused by climatic changes and could be used directly for climate spectral analysis.

An example of the type of automated whole-core logging system that is required is the P-Wave-Logger (PWL) developed at the Institute of Oceangraphic Sciences Deacon Laboratory (Schultheiss and McPhail, in press) for measuring the compressional wave velocity of sediments. The PWL (Fig. 6) accurately measures and automatically records the P-wave velocity of soft sediments within a cylindrical plastic core liner and produces a very detailed velocity log of the whole core.

Velocity measurements are automatically taken at regular intervals along each core section as it travels between a pair of ultrasonic transducers. The PWL provides fine-scale velocity profiles that (a) enable accurate correlations of adjacent cores or holes to be made, (b) provide high-quality data for synthetic seismograms, (c) help in the interpretation of seismic records and (d) indicate the nature of sedimentary features not easily detected by conventional means. The mechanical arrangement of the transducers used on the PWL is shown schematically as part of the overall system diagram in Fig. 6a. An example of the type of data obtained from very high resolution P-Wave logging is shown in Figure 6b which is taken from the *ODP Leg 108 Initial Report* (Schultheiss and McPhail, in press), clearly illustrating the cyclic nature of the sedimentary sequence.

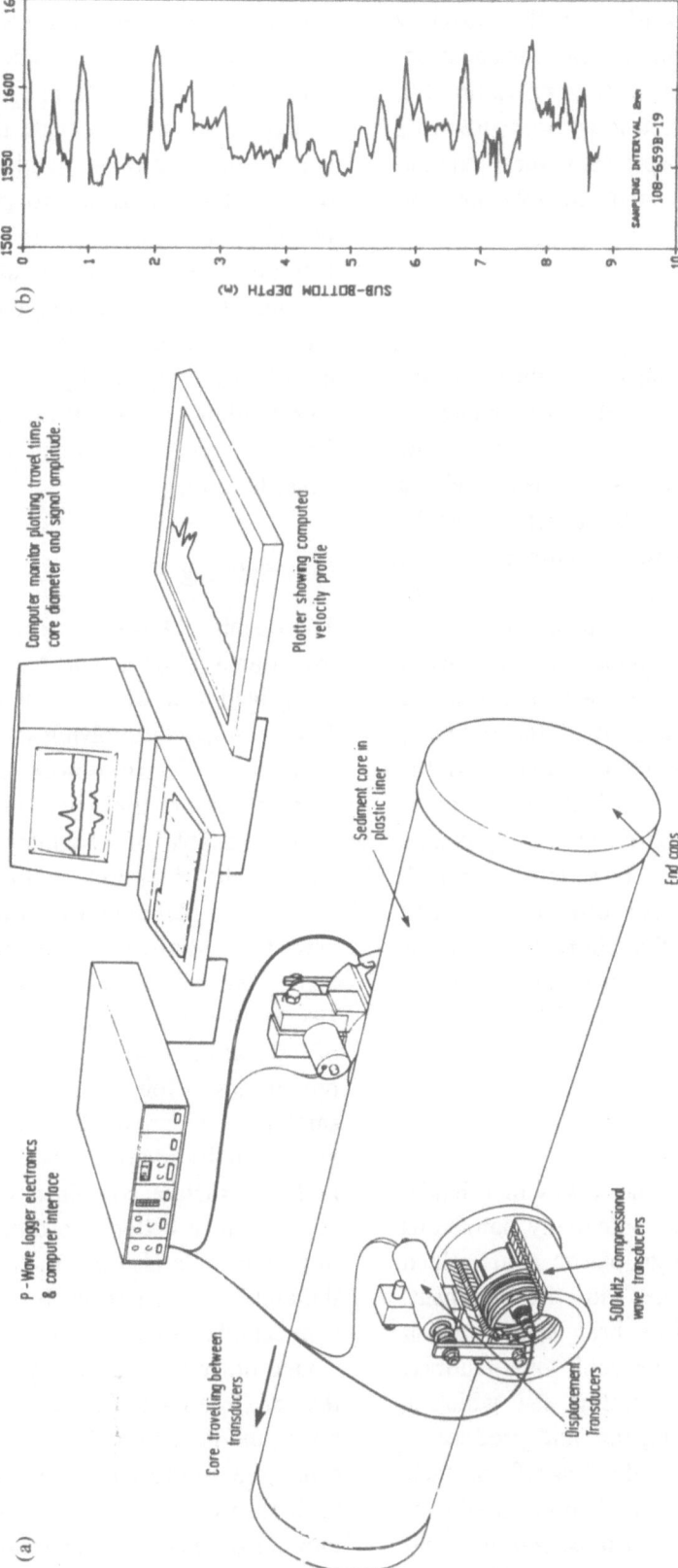

Fig. 6. (a) Schematic diagram of the P-Wave-Logger (PWL); (b) an example of a detailed P-Wave velocity record (after Schultheiss and McPhail, in press).

Currently there is scope for the development of a multi-sensor whole-core logging apparatus that, in addition to providing a tool for the currently established techniques (gamma ray attenuation, P-Wave velocity and magnetic susceptibility), would include other non-destructive measurements such as electrical resistivity, spectral natural gamma, radar, ultrasonic reflection and possibly neutron activation.

Subcoring

Box cores retrieve large volumes of sediment which are often subcored as soon as the core arrives on deck. The IOS box corer is fitted with a vertical row of sample ports which can be easily opened, giving access to a series of sediment layers which can be sampled with syringes before the core is opened (Peters *et al.*, 1980). If the upper closing spade of the box corer is removed the sediment surface will be revealed which can be sampled by pushing in lengths of core liner. Better cores will be taken if a piston is used which can be held at the sediment surface with a small jig which is attached to the corer frame. The box corer described by Papucci *et al.* (1986) has a cylindrical barrel with a segmented liner thus allowing the core to be serially sectioned. For geological investigations it is often important to transport the whole core to the laboratory, and this may be done by using a plastic liner as described by Karl (1976), or by using interchangeable boxes as described by Bouma (1969).

Core Cutting

Core liners can be cut with either a saw or a blade. Rotating saws are noisy and potentially dangerous and produce large amounts of plastic swarf which frequently becomes embedded in the sediment. Other methods have therefore been sought which overcome these problems. Kawohl and Kudrass (1987) report the use of a vibrating saw which is less dangerous than a rotating saw and produces a curled thread of plastic instead of swarf. In their system two pairs of saws are used to cut a slice of sediment (Fig. 7) which can then be X-rayed and sampled, leaving two intact strips of core for storage. This system has the advantage that the core does not require such heavy equipment to hold

it in place during cutting as it does with a blade system.

Core cutters which use blades operate by pulling two blades (one on each side of the core) along the length of the core. It has been found that sharp blades do not work because they cannot be made to cut absolutely straight, and once the cut has gone off line it is impossible to prevent the blade from breaking. The blades are therefore filed flat or broken, and this blunt edge is used. The force required to pull two blades through PVC or polycarbonate liner is considerable, and readily bends the liner if it is not held very rigidly. This requires very strong core holding equipment and it has been found that 1 m-long cores are much easier to handle than 1.5 m-long ones.

Core Splitting

The opening of stored cores is extensively discussed by Bouma (1969). The techniques are all simple, using blades or wires to cut the sediment, although Bouma suggests applying a DC electric current to achieve an electro-osmotic effect if possible. The present authors use a constant current DC power-supply capable of providing 0.5 amp at 30 volts, and attach the cathode to the cutting blade and the anode to a platinum wire which is kept in contact with the sediment and in close proximity to the blade during cutting (Weaver and Schultheiss, 1983b). Sediment surfaces cut in such a way are free of smearing, show burrows and lithological boundaries in minute detail (Fig. 8) and can reveal small-scale sediment disturbances due to coring. A good quality colour photograph of an osmotic knife-cut surface combined with a description produces a much better record of a core than a description alone. We also use the osmotic knife to cut off the archive corner from the Kastenlot cores, and to trim samples for geotechnical testing.

One other method of core splitting is to fracture the core in a controlled way. Fractured surfaces can reveal the maximum detail within the sediments because the surfaces have not been distorted due to local shearing as with a cut surface. This method is unpopular because of the semi-random split which is achieved and the poor visual appearance of the surface. It may be necessary to photograph the core, thus producing a two-dimensional image, before all

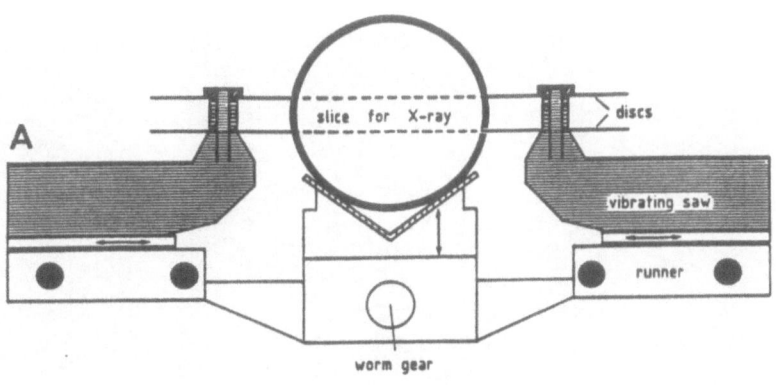

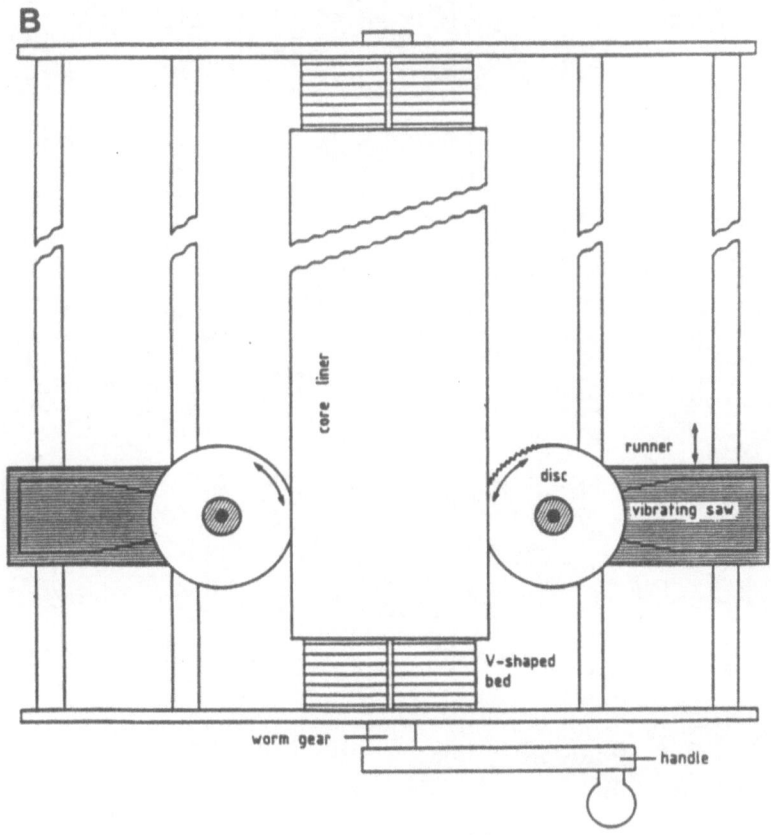

Fig. 7. Vibrating saw for splitting sediment cores into three sections. The centre section is used for X-radiography (from Kawohl and Kudrass, 1987).

the features are readily apparent. An accidental longitudinal core fracture was produced in a Kastenlot core by Weaver and Schultheiss (1983a) which revealed numerous minute open burrows each about 0.5 mm across. Such features would have been virtually impossible to see in a cut surface even with only small amounts of smearing.

Core Description and Photography

Cores are usually described immediately after splitting before any deterioration of the sediment surface occurs. Traditionally, the sedimentologist records all visible features of the core, such as colour, texture, lithological boundaries, burrows etc.

Fig. 8. Examples of heavily burrowed sediment showing the clarity of detail produced by an electro-osmotic knife-cut surface.

The combination of a good quality colour photograph from an unsmeared osmotic knife-cut surface with a multiple whole core log of the core can provide much of the information in a more objective way than from a simple description. Ideally, the photographs and whole core logs should be backed-up by comments from an experienced sedimentologist, and the whole must be synthesized into diagramatic form for publication. One advantage of this method is that information can be interpreted

directly from the core, even if the core is not available, without having to also interpret the potentially poor core description with its mis-interpretations or omissions.

Conclusions

The present range of available corers includes box corers, gravity corers, piston corers, giant piston corers and vibrocorers. If care is taken in selecting the corer most appropriate for the sediment type and most appropriate for the analyses to be conducted, a maximum of 20–30 m of core should be obtainable. Box corers provide good quality cores with large volumes from the upper 0.5 m and often retrieve the sediment–water interface. Single-spade box corers are more successful in obtaining cores in sandy substrates. Gravity corers should be avoided, or cores from them treated with caution for physical property studies or any work which requires precise data on the depths of boundaries within the sediment. This is because they are subject to errors of mis-sampling which cannot always be assessed. The one exception may be the Kastenlot corer, which, because of its large cross-sectional area gives relatively undisturbed cores in soft sediments. Piston corers produce more representative results than gravity corers, but even these are subject to mis-sampling errors due to movement of the piston during the coring operation. These corers can penetrate up to 20 m in suitable sediments. Giant piston corers are capable of coring to 30–50 m. At present, the STACOR is operational and functions well, retrieving relatively undisturbed cores. The GPC has not completed its design phase and so its ability to recover long undisturbed cores is currently unknown.

New developments in corer handling include the use of Kevlar cables which allow more precise monitoring of operations and the production of a containerized winch which allows access to a much larger number of ships for running coring operations.

Whole-core logging techniques combined with colour photographs of osmotic knife-cut surfaces of split cores provide excellent and permanent records of the core which are of more value than descriptions on their own. The authors strongly commend the adoption of these techniques by other workers in the field.

References

Ardus, D. A., Skinner, A., Owens, R., and Pheasant, J., 1982, Improved Coring Techniques and Offshore Laboratory Procedures in Sampling and Shallow Drilling, *Oceanology International* **2**, 18 pp.

Bouma, A. H., 1969, *Methods for the Study of Sedimentary Structures*. J. Wiley & Sons, New York, 458.

Brooke, J. and Pelletier, B. R., 1970, Sea Drilling Techniques of the Bedford Institute, *J. Underwater Sci. and Technol.* **2**, 165–167.

Burns, R. E., 1962, A Note on Some Possible Misinformation from Cores Obtained by Piston-Type Coring Devices, *J. Sedim. Petrol.* **33**, 950–952.

Driscoll, A. H. 1981, The Long Coring Facility, New Techniques in Deep Ocean Coring, *Oceans 81* **1**, New York, IEEE. Inc., 404–410.

Driscoll, A. H. and Hollister, C. D., 1974, The W.H.O.I. Giant Piston Corer; State of the Art, in *Marine Technology Society*, 10th Annual Conference, 663–675.

Driscoll, A. H. and Silva, A. J., 1977, *Report of the Engineering Workshop on Deep Sea Coring*, Vols 1 and 2. Wood's Hole Oceanographic Inst., November, 1977.

Dzwilewzki, P. T. and Driscoll, A. H., 1980, Long Core Facility Winch and Cable System, *American Society of Mechanical Engineers*, Winter Annual Meeting, 7 pp.

Emery, K. O. and Dietz, R. S., 1941, Gravity Coring Instrument and Mechanics of Sediment Coring, *Bull. Geol. Soc. Amer.* **52**, 1685–1714.

Hollister, C. D., Silva, A. J., and Driscoll, A. H., 1973, A Giant Piston Corer, *Ocean Engineering* **2**, 159–168.

Houbolt, J. J. H. C., 1971, Transferable Deep-Sea Coring Gear, *Marine Geol.* **10**, 121–131.

Hvorslev, M. J. and Stetson, H. C., 1946, Free-fall Coring Tube: A New Type of Gravity Bottom Sampler, *Bull. Geol. Soc. Amer.* **57**, 935–950.

Igarashi, Y., Ridlon, J. B., Campbell, J. R., and Allman, R. L., 1970, Note on a Mode of Piston Core Disturbance, *J. Sedim. Petrol.* **40**, 1351–1355.

Karl, H. A., 1976, Box Core Liner System Developed at the Sedimentology Research Laboratory, University of Southern California, *Mar. Geol.* **20**, M1–M6.

Kawohl, H. and Kudras, H. R., 1987, The Use of a Multiple-disc Vibrating Saw for Cutting the Liners of Sediment Cores, *J. Sed. Pet.* **57**, 789–790.

Kermabon, A. and Cortis, U., 1968, A Recoilless Piston for the SACLANTCEN Sphincter Corer, *Saclant ASW Research Centre, Technical Report No. 112*, 22 pp.

Kogler, F. C., 1963, Das Kastenlot, *Meyniana* **13**, 1–7.

Kuijpers, A., Rispens, F. B., and Burger, A. W., 1984, Late Quaternary Sedimentation and Sedimentary Processes of the Madeira Abyssal Plain, Eastern North Atlantic, *Meded. Rijks Geol. Dienst* **38–2**, 91–118.

Kullenberg, B., 1947, The Piston Core Sampler, *Svenska Hydrogr. Biol. Kommn. Skr* **1**, 46 pp.

Kullenberg, B., 1955, A New Core-Sampler, *K. Vet. O. Vitterh. Samh. Handl.* **6**, 17 pp.

Lee, H. J. and Clausner, J. E., 1979, *Seafloor Soil Sampling and Geotechnical Parameter Determination—Handbook, Technical Report*, Civil Engineering Laboratory, Port Hueneme, California **TR-873**, 128 pp.

McCoy, F. W., 1980, Photographic Analysis of Coring, *Mar. Geol* **38**, 263–282.

McCoy, F. W., 1985, Mid-Core Flow-In: Implications for Stretched Stratigraphic Sections in Piston Cores, *J. Sedim. Pet.* **55**, 608–610.

Montarges, R., Fay, J-B., and Le Tirant, P., 1987, Soil Reconnaissance at Great Water Depth, *4th International Conference on Deep Offshore Technology*, 2/18–2/29.

Montarges, R., Le Tirant, P., Wannesson, J., Valery, P., and Berthon, J-L., 1983, Large-size Stationary Piston Corer, *2nd International Conference on Deep Offshore Technology*, 63–74.

Moore, D. G., 1961, The Free Corer: Sediment Sampling without Wire and Winch, *J. Sed. Petrol.* **31**, 627–630.

Moore, T. C. and Heath, G. R., 1978, Sea-floor Sampling Techniques, in J. P. Riley and R. Chester (eds.), *Chemical Oceanography*, 7, Academic Press (London), 75–126.

Papucci, C., Jennings, C. D., and Lavarello, O., 1986, A Modified Box Corer and Extruder for Marine Pollution Studies, *Continental Shelf Research* 6, 671–675.

Peters, R. D., Timmins, N. T., Calvert, S. E., and Morris, R. J., 1980, The IOS Box Corer: Its Design, Development, Operation and Sampling, *I.O.S. Report No. 106*, 16 pp (unpublished manuscript).

Pheasant, J., 1984, A Microprocessor Controlled Seabed Rockdrill/Vibrocorer, *Underwater Technology* 10, 10–14.

Richards, A. F. and Keller, G. H., 1961, A Plastic-Barrel Sediment Corer, *Deep-Sea Research* 8, 306–312.

Schilling, J., Van Weering, T. C. E., and Eisma, D., 1988, Advantages of Lightweight Kevlar Rope for Ocean Bottom Sampling with Piston Corer and Box Corer, *Mar. Geol.* 79, 149–152.

Schultheiss, P. J. and Mienert, J., 1988, Whole Core P-Wave Velocity and Gamma Ray Attenuation Logs from ODP Leg 108 (Sites 657–668), in W. Ruddiman, M. Sarnthein, J. Saldauf *et al.*, *Proc., Init. Repts. (Pt. A), ODP* 108, 1015–1046.

Schultheiss, P. J. and McPhail, S. D., 1989, An Automated P-Wave Logger for Recording Fine Scale Compressional Wave Velocities in Sediments, in W. Ruddiman, M. Sarnthein, J. Baldauf *et al.*, *Proc., Init. Repts. (Pt. B), ODP* 108 (in press).

Schuttenhelm, R. T. E., Auffret, G. A., Buckley, D. E., Cranston, R. E., Murray, C. N., Sheppard, L. E., and Spijkstra, A. E. (eds). 1990, *Geoscience Investigations of Two North Atlantic Abyssal Plains— the ESOPE International Expedition*, Vols 1 and 2, CEC-Joint Research Centre, JRC Report (in press).

Silva, A. J., Hollister, C. D., Laine, E. P., and Beverly, B. E., 1977, Geotechnical Properties of Deep-Sea Sediments: Bermuda Rise, *Mar. Geotechnol.* 1, 195–232.

Weaver, P. P. E. and Schultheiss, P. J., 1983a, Vertical Open Burrows in Deep Sea Sediments 2 m in Length, *Nature* 301, 329–331.

Weaver, P. P. E. and Schultheiss, P. J., 1983b, Detection of Repenetration and Sediment Disturbance in Open Barrel Gravity Cores, *J. Sedim. Petrol* 53, 649–654.

Observation of Corer Penetration and Sample Entry during Gravity Coring

W. R. PARKER and G. C. SILLS

Blackdown Consultants, Taunton and Department of Engineering Science, Oxford University, UK

(Received 27 April, 1989; 1 September, 1989)

Key words: gravity coring, core-shortening, corer penetration, sample entry, observations.

Abstract. Gravity core samples provide the basic data source for a wide range of geological, geotechnical and geochemical studies. However, the length of the core recovered is often less than the penetration achieved by the corer, such cores being described as "shortened". If the penetration of the corer has been measured, and it is assumed that no dropout of the core occurs as the barrel is withdrawn from the seabed, it is present practice to reconstruct *in situ* dimensions using an overall correction factor based on this penetration and the length of core recovered. However, measurements, reported here, have been made of corer penetration and sample entry and these show that the entry deficit (penetration minus sample entry) develops in some instances continuously, and in others intermittently. These results indicate that an overall correction factor is unlikely to be appropriate to any given section of the core.

Background

A number of authors (e.g. Emery and Deitz, 1941; Hvorslev, 1949; Emery and Hulsemann, 1964; Ross and Riedel, 1967; McCoy, 1971; Lebel *et al.*, 1982; Weaver and Schultheiss, 1983) have reported observations made during the process of gravity coring that the recovered core length is frequently less than the depth of penetration of the corer into the sediment. McCoy (1971), Lebel *et al.*, (1982) and Weaver and Schultheiss (1983) have discussed the causes of this phenomenon and have investigated its nature by a variety of comparative observations.

Although penetration of the corer has sometimes been measured during the coring process, the corresponding measurement of the sample entering the corer has not, in general, been made. Sample lengths can be measured after recovery, but these values may be affected by the further process of "dropout" from the corer as it is pulled from the seabed. This chapter described measurements of penetration of the corer into the bed and of entry of the sample into the corer, made during coring. Any losses that occur at this stage will be crucial in subsequent analyses which rely on a knowledge of depth of the sample in the seabed.

Equipment and Methods

A system to measure corer penetration and sample entry was developed using two acoustic transducers mounted on an adaptor fitted between the core barrel and corer weight chassis. One transducer was directed parallel to the barrel outside the corer to the seabed and the other inside the core barrel to the surface of the sample. The transit time from sending a signal to receiving its reflection from the sediment was measured with a dual channel echo-ranging system. The time interval was converted to distance by measurement of the speed of sound in water, using an external target at a known distance on the outside of the corer barrel and the core catcher as a internal target. An analogue DC output proportional to distance was recorded onboard ship. The operating frequency of the transducers was chosen to be centred around 500 kHz, and the pulse repetition rate was approximately 400 Hz. The system allows continuous monitoring of the corer penetration and withdrawal, the sample entry and, if it occurs, subsequent dropout. Under field conditions the overall resolution of the system is 5 mm.

Two types of corers were studied: first a cylindrical barrel gravity corer with liner I.D. = 83 mm, barrel O.D. = 101 mm and barrel lengths of 3.17 and 3.30 m; and secondly a 150 mm × 150 mm square section Kasten corer with a 2.3 m barrel. The area ratio (Hvorslev 1949, Emery and Hulsemann 1964) of the cylindrical barrel corer was 0.33 and that of the Kasten corer was 0.05, values which are typical for such corers.

Sites

Cores were collected at five sites in the northeast Irish Sea. Detailed observations are presented for two sites, IS05 at 54° 24′ N, 3° 41′ W and ISO 1 at 54° N 3° 39′W. The sediment of the area is a silty sand mud (Pantin, 1978, Williams *et al.*, 1981), which has been heavily bioturbated to considerable depth (Kershaw *et al.*, 1983, Kershaw *et al.*, 1984). X-radiographs of cores show that generally it is of homogeneous texture (Williams *et al.* 1981; Kirby *et al.* 1983). At each site, seabed density profiles were measured *in situ* using a nuclear backscatter density probe (Parker *et al.* 1975). Since the sediment is contaminated by discharges of low-level radioactive waste from Sellafield, the results were corrected for background radiation. Density in core subsamples was measured onboard using a vibrating tube densimeter (that converts from natural frequency of vibration to density) and, in the laboratory ashore, by moisture content determination. These data taken together suggest a similar pattern at both sites. Densities increase downwards in the top 0.4 m of the bed to approximately 1.7 Mg/m^3, then increase more slowly to about 1.9 Mg/m^3 by 1.4 m, thereafter remaining more or less constant to 2 m. Shear strength was measured on core samples to a depth of 2 m using a standard vane and sensitive torque transducer. Results show strength increasing with depth,

with peak values of the order of 10 kN/m^2 and residual values of 2 kN/m^2 reached by a depth of 2 m.

Data Analysis

Changes in distance to the sediment surface outside the corer give approach speed and penetration, P, and inside the core barrel yield sample entry, E. In this analysis core entry was determined from the chart records at each 100 mm increment in penetration. Core entry deficit (P-E) and core entry ratio (E/P) were thus determined throughout corer penetration. The incremental entry deficit, which represents the entry deficit arising in each 100 mm of penetration, and the incremental entry ratio (the ratio of entry to penetration in each penetration increment) were also calculated. Increases in range from the external echo sounder allow withdrawal to be monitored and increases in range from the internal sounder indicated sample dropout. These values can be correlated with recovered core lengths.

Results

Core entry ratios have been calculated for 34 deployments of a 3 m-long cylindrical barrel corer and 14 deployments of a 2 m-long Kasten corer. The results are plotted against approach speed in Fig. 1. The groupings of the data reflect available winch speeds, but the actual approach speed was derived from the chart record to an accuracy of 0.05 m/sec. The entry ratios vary non-systematically from one core to

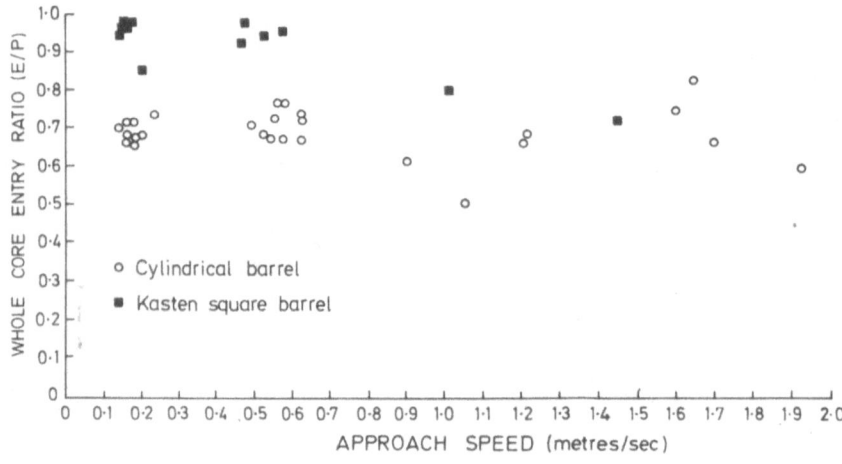

Fig. 1. Calculated core entry ratios for corer types.

another, even when speeds of entry are similar. The cylindrical barrel cores show no significant trend in entry ratio for approach speeds up to 1.9 m/sec, although the range of values increases at the higher speeds. If the core entry ratios are calculated for the barrel cores for the first 2 m of penetration (to make them directly comparable with the Kasten core results), the individual values change, some increasing and some decreasing, by amounts generally up to about 0.04, but the overall pattern of variation remains very similar. At approach speeds below 0.6 m/sec, higher entry ratios are achieved with the Kasten corer than with the cylindrical barrel corer. At speeds above 1.0 m/sec, the entry ratio in the Kasten cores declines to that for the cylindrical barrel cores.

ENTRY DEFICIT AND ENTRY RATIO PROFILES

Profiles of cumulative and incremental entry deficit *vs* total penetration for the barrel corer are shown in Figs 2 and 3. The cores illustrated were taken in a period of 2 hours with the ship at anchor at site IS05. Only the approach speed varied and the profiles are arranged in order of increasing speed. Although the cumulative entry deficit increases steadily in cores 44 and 45 (Fig. 2), the incremental deficit is irregular. This is more clearly shown in core 47 (Fig. 3a). An extreme example is shown in core 46 (Fig. 3b), where adjacent sections of core show incremental deficits of zero (complete entry locally) and 60 mm (in 100 mm penetration only 40 mm entered). Also indicated in Figs 2 and 3 are the whole core entry ratio and the incremental entry ratio (scale from right to left) clearly illustrating the local variability in entry ratio compared to the whole core value shown.

Cumulative and incremental entry deficit profiles for four Kasten cores from site IS01 are plotted in Fig. 4. Also marked are the whole-core entry ratio and incremental entry ratio scale. Only the approach speed was varied and the profiles are arranged in order of increasing speed. The profiles for cores 68 and 78 are typical of the high entry ratio achieved. Cores 76 and 73 illustrated the values of direct continuous observation by revealing the distribution of entry deficit. Core 73 shows significant deficit growth in the upper two thirds of the core with none below 1.6 m, whereas core 76 shows most of the entry deficit arising during the latter 1 m of penetra-

tion. The differences between whole-core entry ratio and local entry ratio are also clearly visible.

Discussion

Previous authors (e.g. Emery and Deitz, 1941; Hvorslev, 1949; Emery and Hulsemann, 1964; McCoy, 1971; Label *et al.*, 1982 and Karnes *et al.*, 1980) have suggested that approach speed influences entry ratio in barrel corer operation. For the barrel corer tested and at the approach speeds used these data show little systematic variation in entry ratio. The Kasten corer tested provided cores with generally good entry ratio except at high approach-speed. Several authors have suggested that entry deficit develops linearly (Emery and Deitz, 1941; Emery and Hulsemann, 1964; and Lebel *et al.*, 1982) or increases with penetration (Ross and Riedel, 1967; Weaver and Schultheiss, 1983) and that the entry ratio for the whole core is representative of that for a particular level (Emery and Hulsemann, 1964; Lebel *et al.*, 1982). Data illustrated in Figs 2–4 show clearly that whole-core values mask the irregularity in the distributions of entry deficit or entry ratio which do not necessarily develop either continuously, linearly or increasingly with depth. For example, in Fig. 2a, the incremental entry deficit is distributed around the whole-core entry ratio over the whole depth of penetration, while Fig. 2b shows the incremental entry deficit to be generally lower than the whole-core entry ratio in the upper 2 m of the core, and higher in the metre below.

It has been suggested that sample losses may be due to compression. However, on the timescale of corer penetration (10–60 sec) compression of the sediment framework requiring expulsion of pore water is not feasible for the type of sediment sampled. The difference between penetration and sample length must therefore arise as a consequence of sediment not entering the core barrel but being pushed aside to a greater or lesser degree with, at times, no sediment entering. It is possible that material may be carried down ahead of the corer to enter later when the balance of forces influencing sample entry changes. Thus, the entry deficit at the particular level of penetration does not necessarily indicate the absence of the stratigraphic level in the core. Cores which are shorter than corer penetra-

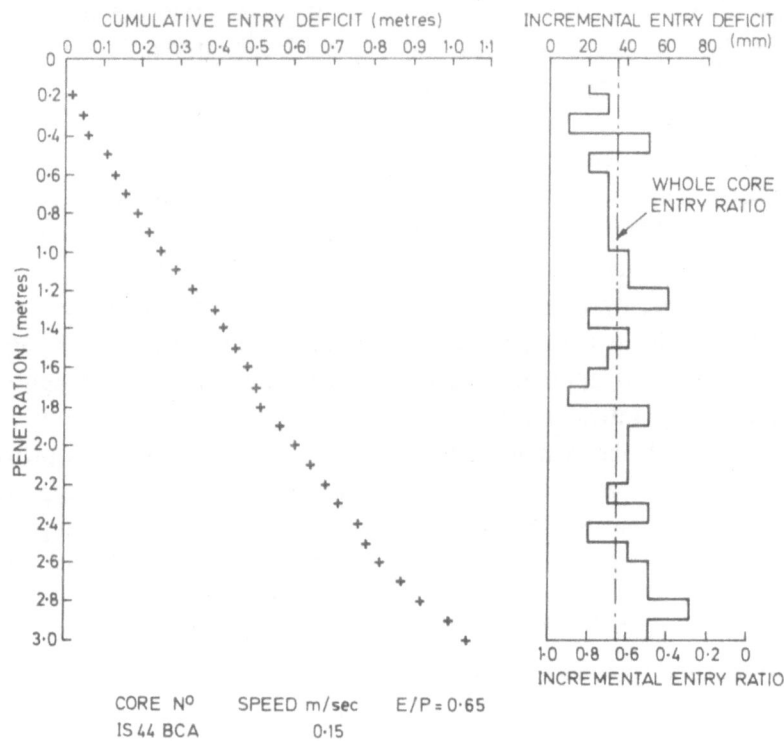

Fig. 2(a) Cumulative and incremental entry deficit profiles: barrel core 44.

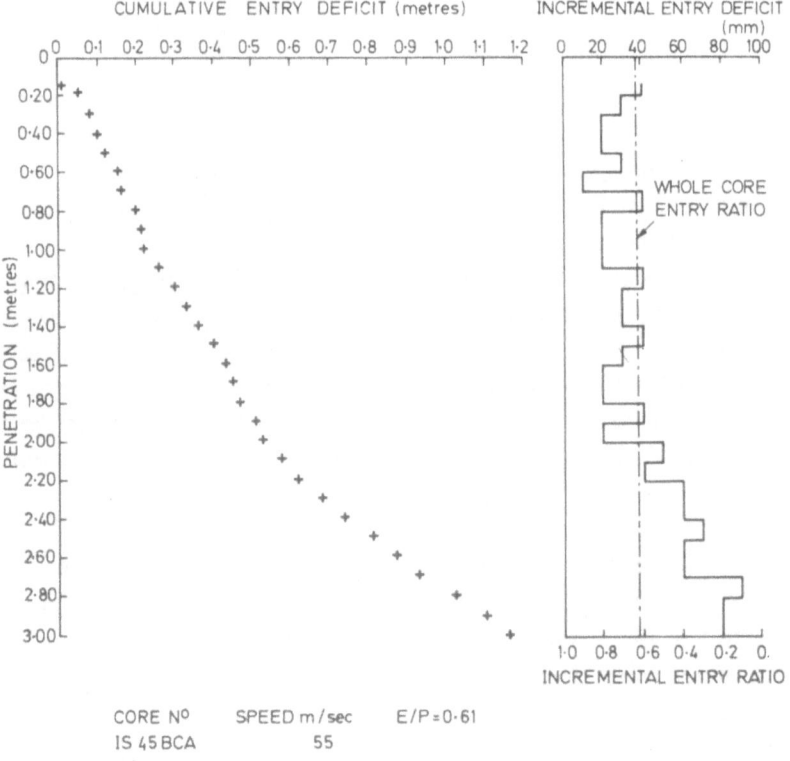

Fig. 2(b) Cumulative and incremental entry deficit profiles: barrel core 45.

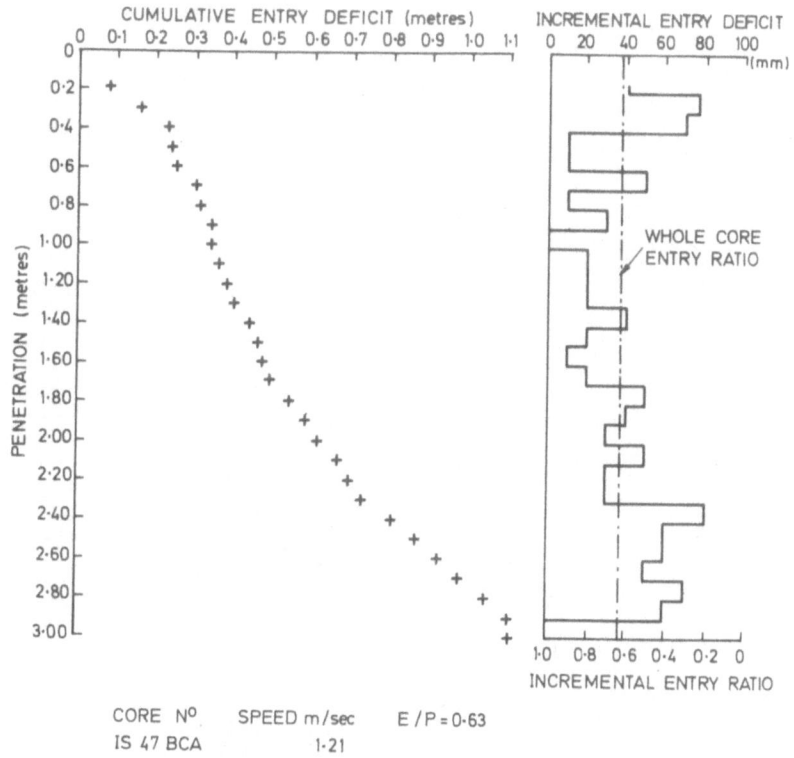

Fig. 3(a) Cumulative and incremental entry deficit profiles: barrel core 47.

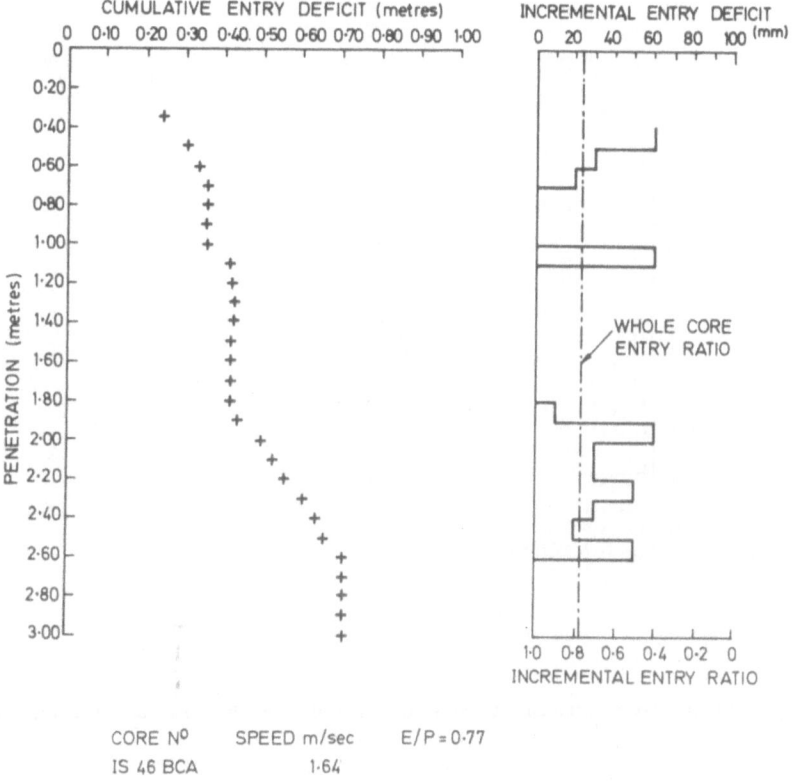

Fig. 3(b) Cumulative and incremental entry deficit profiles: barrel core 46.

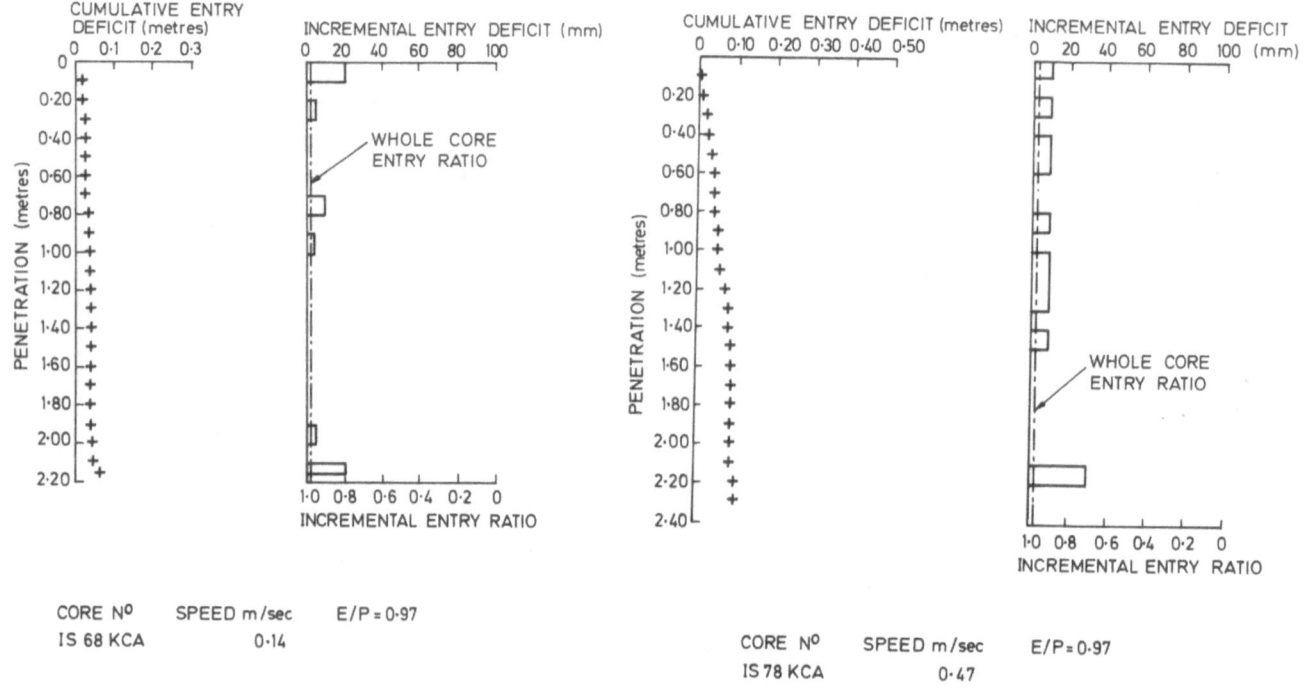

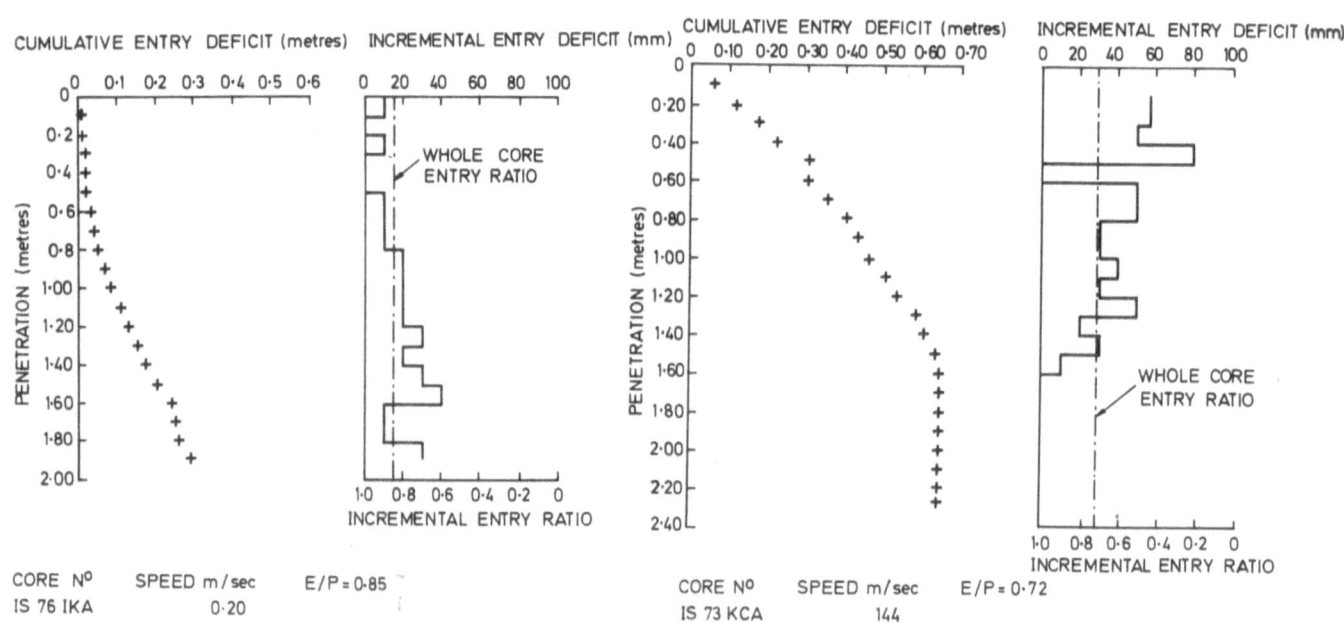

Fig. 4. Cumulative and incremental entry deficit profiles for four Kaster cores: site ISO 1.

tion can occur as a result of entry decifit of "dropout". The description "core-shortening" sustains the concept of overall volume reduction during coring which is a misconception of the process involved. Ideally, data derived from gravity cores should be accompanied by an entry deficit or entry ratio profile.

Conclusions

The Kasten corer produces substantially less entry deficient samples than the barrel corer. This is most probably due to the difference in area ratio between the two corers. For the cylindrical barrel equipment and sediment type reported here whole-core entry ratios vary non-systematically from 0.50 to 0.82 for a range of approach speeds up to 1.9 m/sec. The incremental entry ratio varies with penetration and may be significantly larger or smaller than the whole-core entry ratio. Thus, whole-core ratios are not a guide to sample deficiency at particular levels within a core and their use in "linear stretching" of core profiles may compound existing deficiencies. Entry deficit profiles do not allow complete reconstruction of the position of individual sections of core. However, they do identify those parts of the penetration event during which entry of the sample is deficient or absent and thereby indicate the quality of the core for studies which rely on its dimensional integrity. These results emphasize the need to choose a corer which generates minimum entry deficit and demonstrate the desirability of measuring the pattern of entry deficit for each core.

Acknowledgements

This work was supported by the Department of the Environment under contract No. PECD/7/9/172-150-83 and by the Ministry of Agriculture, Fisheries and Food. The contribution of colleagues in Oxford, and in Lowestoft, and the officers and crew of R.V. Cirolana is gratefully acknowledged. Special thanks are due to Dr. D. McG. Elder, currently with Soils and Foundations Ltd, New Zealand, and previously at Oxford, who provided the shear strength and density data.

References

Emery, K. O. and Hulsemann, J., 1964, Shortening of Sediment Cores Collected in Open Barrel Gravity Corers, *Sedimentology* **3**, 144–154.

Emery, K. O. and Deitz, R. S., 1941, Gravity Coring Instrument and Mechanics of Sediment Coring, *Bull. Geol. Soc. Amer.* **52**, 1685–1714.

Hvorslev, M. J., 1949, *Subsurface Exploration and Sampling of Soils for Civil Engineering Purposes*, US Corps Engs. Waterways Expt. Sta. Tech. Repts.

Karnes, C. H., Burchett, S. N., and Dzwilewski, P. T., 1980, *Optimised Design and Predicted Performance of a Deep Ocean 50 m Piston Coring System*, I.E.E.E. Publication No. 80CH1572-7, pp. 231–239.

Kershaw, P. J., Swift, D. J., Pentreath, R. J., and Lovett, M. B., 1983, Plutonium Redistribution by Biological Activity in Irish Sea Sediments, *Nature* **306**, 774–775.

Kershaw, P. J., Swift, D. J., Pentreath, R. J., and Lovett, M. B., 1984, *Science of the Total Environment* **40**, 61–81.

Kirby, R. Parker, W. R., Pentreath, R. J., and Lovett, M. B. 1983, *Sedimentation Studies Relevant to Low-level Radioactive Effluent Dispersal in the Irish Sea*, Rep. No. 178, Institute of Oceanographic Sciences, Taunton.

Lebel, J., Silverberg, N., and Sundby, B., 1982, Gravity Core Shortening and Pore Water Chemical Gradients, *Deep Sea Research* **29**, 1365–1372.

McCoy, P. W., 1971, An Analysis of Piston Coring through Corehead Camera Photography, *Underwater Soil Sampling, Testing and Construction Control*, A.S.T.M., STP **501**, 90–105.

Pantin, H. M., 1978, Quaternary Sediments from the North-east Irish Sea: Isle of Man to Cumbria, *Bulletin of the Geological Survey of Great Britain*, No. 64.

Parker, W. R., Sills, G. C., and Paske, R. A. E., 1975, *In situ Bulk Density Measurement in Dredging Practice and Control*, *First Intl. Symp. Dredging Technology*, *BHRA*, Cranfield.

Ross, D. A. and Riedel, W. R., 1967, Comparison of Upper Parts of some Piston Cores with Simultaneously Collected Open-Barrel Cores. *Deep Sea Research* **14**, 285–294.

Weaver, P. P. E. and Schultheiss, P. J., 1983, Detection of Repenetration and Sediment Disturbance in Open-Barrel Gravity Cores, *Jour. Sed. Pet.* **53**(2), 649–654.

Williams, S. J., Kirby, R., Smith, T. J., and Parker, W. R., 1981, *Sedimentation Studies Relevant to Low-Level Radioactive Effluent Dispersal in the Irish Sea. (Part II)*, Rep. No. 120, Institute of Oceanographic Sciences, Taunton.

Ocean Drilling Program (ODP) Deep Sea Coring Techniques*

M. A. STORMS

Ocean Drilling Program, Texas A & M University, U.S.A.

(Received 27 April, 1989; accepted 1 September, 1989)

Key words: coring, wireline, scientific, ocean, downhole, deepwater Resolution, Challenger, ODP, DSDP.

Abstract. The coring techniques and systems of the Ocean Drilling Program (ODP) were developed to satisfy a scientific need for better quality and improved recovery of oceanic core samples. Some of the ODP systems in use today evolved from refinements to earlier systems developed by the Deep Sea Drilling Project (DSDP) or were adaptions of available industry technology. Other systems were conceived and designed by ODP engineers. The evolution of these progressive scientific coring systems began with the Rotary Core Barrel (RCB) used by the DSDP and proceeded to the highly advanced Diamond Coring System (DCS) currently under development by ODP engineers. During the evolution several other key systems were developed. These included hydraulic piston coring, extended coring, pressure coring, bare rock spudding and several systems using high speed diamond coring technology. These include the positive displacement coring motor (PDCM), the "Navi-drill" core barrel, and a top driven diamond coring system (DCS). This article describes the evolution and conceptual design of these systems including the required bottom hole assemblies, and the sinker bar/sandline configurations.

Historical Background

The coring systems in use today by the Ocean Drilling Program (ODP) evolved both as refinements to earlier systems developed by the Deep Sea Drilling Project (DSDP) and because of a continual need by the marine geology community as a whole. In general, the coring systems were developed from unique, internally generated, concepts or were adapted from existing industry systems to satisfy a scientific need for better quality and improved recovery of oceanic core samples.

In 1968, the National Science Foundation-funded (DSDP) embarked on an eighteen-month scientific mission to recover marine geology core samples by drilling a series of holes (transect) across the Mid Atlantic Ridge. The intent of this research program, managed by the University of California, San Diego, Scripp's Institution of Oceanography, was to prove or disprove once and for all the theory of sea floor spreading. The effort utilized a specially designed, technically advanced, dynamically positioned drill ship, called the *Glomar Challenger* (Fig. 1). This drill ship was designed and operated by Global Marine Incorporated (GMI). Using the "Challenger" the DSDP research program successfully recovered the deepest cores ever recovered from an offshore drill ship and successfully proved the theory of sea floor spreading. The predominant coring system used for this drift research effort was a wireline retrievable rotary coring system developed by Hycalog, a prominent "oilfield" service and supply company. Due to the dramatic success of this "pilot" program and the immense scientific potential of continued scientific coring, the project was extended and went on to recover literally miles of deep-sea core samples and over a period of 15 years proved and disproved many marine geological theories and ideas.

The follow-on project to DSDP began in 1983 and was christened "the Ocean Drilling Program" (ODP). The ODP is managed by Texas A & M University and operates the state-of-the-art scientific drill ship *Joides Resolution* registered as the SEDCO/BP 471 (Fig. 2). The principal funding agency for the ODP remains the National Science Foundation; however the program receives significant monetary, scientific and technical support from its international partners which include: the Federal Republic of Germany, France, Japan, the United Kingdom, Canada and Australia and the European Science Foundation Consortium.

Fig. 1. D/V *Glomar Challenger*.

Fig. 2. D/V *Joides* resolution.

The initial coring system used by DSDP was wire-line retrievable and represented a reasonably efficient system for recovering continuous core samples from the deep ocean. The 2.44 in. (6.20 cm) diameter by 32 ft (9.76 m) long core was cut by the rotation of a tungsten carbide insert (TCI) roller cone core bit (see ODP operation in Fig. 5). The core was received in a clear plastic, acetate butyrate core liner installed inside the wireline retrievable core barrel. A collet-type core catcher retained the hard, crystalline rock core and kept it from falling out of the bottom of the barrel during retrieval. The core barrel was equipped with a latch at the top to counteract the torque generated by the coring process. A swivel assembly, installed immediately below the core barrel latch, in conjunction with a lower support bearing installed as a landing shoulder in the outer core barrel assembly, prevented the inner core barrel from rotating during the coring process and thus disturbing the core. A

check valve installed in the swivel allowed water to be displaced out of the top as the core entered the barrel at the bottom. This feature was designed to minimize core erosion and lost recovery. The check valve also prevented the core from being washed out of the core barrel during retrieval back to the drill ship.

Although the coring system was effective initially for recovering hard-rock core, it was not optimum for other types of lithologies, particularly the softer sediments encountered at or near the mudline. The upper 100 to 300 m of the sedimentary column were often highly disturbed by the rotary drilling process. Attempts at detailed sampling for disciplines such as paleoceangraphy, paleoclimatology, magnetostratigraphy, and high-resolution stratigraphy were all but impossible. Initially, improvements were made to the basic rotary coring system such as new spring-loaded dog-type core catchers and more efficient venting at the top of the core barrel to minimize the back pressure that tends to inhibit core entry. These changes helped to improve overall system performance, particularly core recovery, but still fell short of the scientific requirement for good quality, soft formation cores.

It was apparent that some means to overcome the limitations of rotary drilling in unlithified sediments was required. Oceanographic piston cores studies had provided means of distinguishing events recorded in sediments to a precision of thousands of years; such events were being homogenized by DSDP's rotary coring. Although oceanographic vessels were routinely taking piston cores of mudline sediments, these "conventional" piston coring systems were limited to just a few tens of meters of the surface material and lacked the capability to achieve any significant depth of penetration.

The original DSDP hydraulic piston coror (HPC) was born of rotary coring shortcomings. The intention was to develop a high-quality coring system, compatible with development through the drill string and capable of penetrations up to 200 m below the sea floor. The HPC system was developed to meet this mandate. It has since been refined through several iterations to the enormously successful Advanced Piston Corer (APC) system in use today in the ODP (see ODP operation in Fig. 10). Major hydraulic piston coring refinements that have taken place over the years include expanding the core

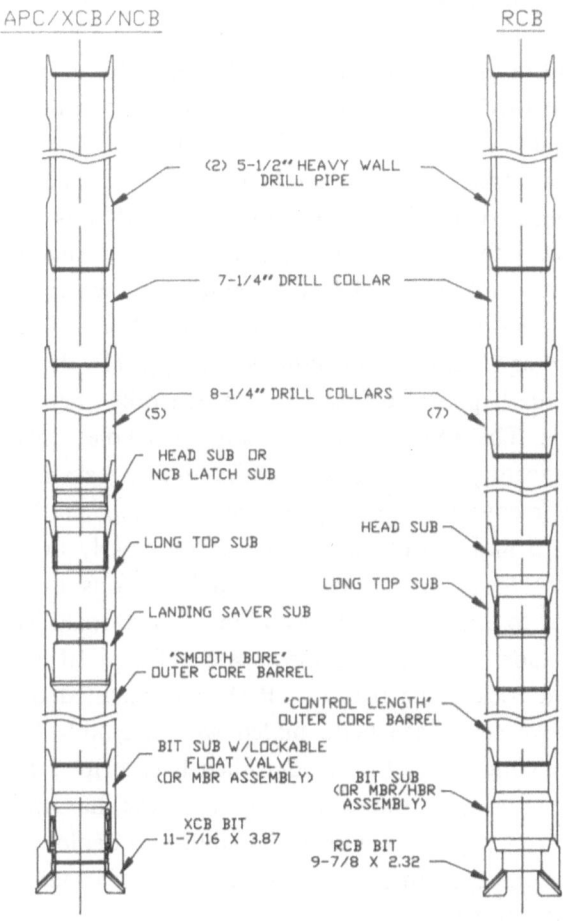

APC/XCB/NCB   RCB

(2) 5-1/2" HEAVY WALL
DRILL PIPE

7-1/4" DRILL COLLAR

8-1/4" DRILL COLLARS
(5)   (7)

HEAD SUB OR
NCB LATCH SUB

LONG TOP SUB   HEAD SUB

LANDING SAVER SUB   LONG TOP SUB

'SMOOTH BORE'
OUTER CORE BARREL

                                       'CONTROL LENGTH'
                                       OUTER CORE BARREL

BIT SUB W/LOCKABLE
FLOAT VALVE
(OR MBR ASSEMBLY)   BIT SUB
                                       (OR MBR/HBR
                                       ASSEMBLY)
XCB BIT
11-7/16 X 3.87

                                       RCB BIT
                                       9-7/8 X 2.32

Fig. 3. Typical ODP bottom hole assembly configurations.

length from the original 14.8 ft (4.5 m) to the present 31.2 ft (9.5 m), increasing thrust to allow coring into stiffer formations, adding a core orientation capability, adding temperature measurement capability, keyed piston rods to eliminate core barrel spiraling, modified shear subs for more consistent shearing pressures, and finally refining all components for more simple and efficient operations.

The current APC system has cored in excess of 300 m below the sea floor (mbsf) and routinely provides 90 to 100% core recovery. It has the capability to penetrate into semi-indurated formations which can then be rotary cored without inducing significant core disturbance. As such, the APC is now the more widely used coring system in the ODP arsenal.

The tremendous success and acceptance of hydraulic piston coring within DSDP led to the development of the extended core barrel (XCB) coring system. With the almost exclusive use of the HPC system for spudding holes a dilemma arose. The

bottom hole assembly required for piston coring operations was not compatible with the rotary coring system. The core bit for HPC coring has the cones spread apart to allow passage of the piston core barrel out into the formation ahead of the bit face. This bit would cut a 3.88 in (9.86 cm) OD core, much too large to fit inside the standard 2.88 in (7.32 cm) ID rotary core barrel. This meant that upon completion of each piston cored hole, the drill string would have to be tripped out and the BHA changed to allow deeper objectives and/or basement to be reached. Obviously this was a very time-consuming operation and was not an efficient way to continuously core in the deep oceans.

The extended core barrel (XCB) was developed to rectify this situation by allowing a "driven" rotary coring system to be deployed in the APC BHA (see ODP operation in Fig. 13). The XCB design required an independent cutting shoe to trim the core from the 3.88 in (9.86 cm) OD size provides from the roller cone bit to a nominal 2.31 in (5.87 cm) OD size that could be recovered inside the core barrel. The development of the XCB system allowed the deepening of single bit HPC holes without requiring the tripping of the drill string and subsequent re-spudding of the hole. As with the HPC, the XCB development has continued through many iterations throughout the years. Enhancements to the system included better circulation to the cutting shoe, strengthened connections for deeper drilling into more indurated formations, the incorporation of a "venturi" venting assembly resulting in further reduction of core barrel back pressure, and many hardware improvements geared toward simpler, more efficient shipboard operations. The current version of the XCB has drilled in excess of 1000 m below the sea floor and has significantly improved recovery and quality in most formations.

Lithologies which still create problems for the XCB are interbedded soft/hard formations (e.g. chalk/chert), unconsolidated material such as turbidities or loose-flowing sands, and basement-type crystalline rock. The ODP is pursuing the evaluation of hydraulic percussion hammer (vibra-coring) techniques as a possible way to improve recovery in unconsolidated formations. Improved basement and crystalline rock coring systems are being developed using small diameter, narrow kerf, high speed, diamond coring techniques.

The development of ODP diamond coring began with the navi-drill core barrel (NCB). The NCB is the third of several coring systems either designed or under development within DSDP/ODP to enable single-bit holes to be drilled with the optimum coring system from the mudline down to and possibly into basement. As with the APC and XCB coring systems the NCB is fully compatible with the APC/XCB BHA. This system is being developed to allow coring to continue once basement or crystalline rock has been reached with the XCB. When fully operational the NCB should allow limited coring (i.e. 10–50 m) into basement.

The most recent ODP coring system currently under development to be compatible with the APC/XCB BHA is the pressure core sampler (PCS). The predecessor to the ODP PCS was the DSDP pressure core barrel (PCB). The DSDP version, although successful at recovering cores under pressure, had several major shortcomings that eventually led to the scientific mandate for an improved pressure coring capability for the ODP. The PCB was developed before the APC revolution. It therefore was only compatible with the rotary coring system (RCB). Today the RCB system is only rarely used for coring softer lithologies and so the PCB was incompatible for most desired applications. The pressure chamber was very long and cumbersome causing special handling difficulties on deck. The PCB operating pressure was limited to 5000 psi (340 bar), which negated it's use in many deep-water situations. Finally, the PCB was extremely complex to fabricate (hence expensive) and was very difficult to assemble and operate.

The new ODP PCS system is designed to recover 61 in.^{3} (1000 cm^3) samples, 1.65 in. (4.19 cm) in diameter and recover them under *in situ* (hydrostatic) pressure. This system has a maximum operating pressure of 10 000 lb/in.^{2} and, as was stated earlier, is fully compatible with the APC/XCB BHA. The PCS is significantly cheaper to fabricate and is much easier to assemble, handle and deploy at sea. When completed, the PCS system will have a special core processing chamber on-deck to receive the pressurized core sample. The chamber will be pressure/temperature compensated, and will have provisions for adapting to other scientific equipment provided by independently funded principal investigators.

Besides sharing total compatibility within the APC/XCB BHA the four aforementioned coring systems (APC/XCB/NCB and PCS) are completely interchangeable at any point in the coring cycle. Any of the coring systems may be substituted for another when necessary to meet the scientific objectives for the particular formation being cored.

Independent of the single bit hole "optimized" coring systems described above, the ODP was also directed to develop the capability to spud (initiate drilling and coring) holes on bare rock, where the basement rock is exposed to the sea floor. These "zero-age" crustal formations are typically highly fractured, poorly cemented, and generally very unstable. The ability to successfully spud holes, make significant penetration, and maintain any reasonable core recovery percentage in this environment had been a major problem for the DSDP. Early in the ODP (November 1985) the capability was established to allow successful spud-in on bare rock. The sea floor structure developed was referred to as a "hard rock guide base" (HRB) and consisted of a massive steel base plate mated with a gimbled re-entry cone above the cement filled bags below. The technique was time-consuming and the hardware relatively expensive, but the capability to deploy the HRB and spud-in on bare rock was demonstrated on ODP legs 106 and 118.

Although the ability to spud-in on fractured rock was proven, achieving significant penetration and core recovery in this environment continued to be a problem for the ODP. This led to a dramatic departure from conventional offshore coring techniques. The development of a coring system designed to allow the deployment of high speed, narrow kerf, diamond core bits from a floating vessel was initiated. This system, dubbed the diamond coring system (DCS), was based on proven diamond coring systems successfully in use by the mining industry for ore body exploration. The DCS system required mounting a smaller, high speed, top drive on to a platform hung from the main 400 ton (363 000 kg) heave compensator in the derrick and then drilling through standard ODP drill pipe with a smaller, nominal 3.5 in (8.89 cm) diameter, DCS work string (tubing). The cores are then wireline retrieved through the work string. Although using proven diamond coring technology (for land drilling), the DCS was highly developmental as an offshore coring system because it required very precise weight on bit control. Whereas conventional offshore coring operations with TCI roller cone bits might require 25 000–35 000 lb (11 350–15 890 kg) Weight On Bit WOB and fluctuate 10 000–15 000 lb (4540–6810 kg) due to vessel motion, the DCS small, 4.0 in (10.16 cm) diameter narrow kerf, bits required 4000–12 000 lb (1810–5440 kg) WOB with fluctuations of plus or minus 500 lb (227 kg). To maintain that level of control for the DCS coring required the development of a secondary "active" heave compensation system. This second compensator utilized load cells, accelerometers and hydraulic servo valves to reduce the residual heave resulting from the large compensator inefficiencies and differential stretch of the two concentric pipe strings. Initially conceived as a coring system for fractured crystalline rock for the DCS has demonstrated great potential for coring in interbedded formations, atoll/guyot coring, shallow water carbonate coring such as coral and reefal limestones, and as a possible high-temperature coring system for future hydrothermal drilling environments.

Future coring systems that may well be in operation some day include a 24 600 ft (7500 m), 5 in (12.7 cm) diameter DCS with the platform requirement eliminated and the main drill string hung off on riser tensioners, a "smart" core barrel that yields real time feedback of core blockage and/or loss of recovery to the driller, and rock coring systems using high pressure water jet technology.

Having described the basic evolution and coring philosophy of the DSDP and ODP scientific coring projects the remainder of this chapter will deal with the design and operation details of each coring system in use or under development including a section on the bottom hole assemblies into which these coring systems are deployed and the wireline sinker bar assemblies with which the coring systems are deployed and/or recovered.

Bottom Hole Assembly

The bottom hole assembly or BHA as it is commonly referred to in industry represents that hardware that is attached below the lower most joint of the primary drill string (Fig. 3). This commonly includes such items as the core bit, outer core barrel assembly, drill collars, stabilizers, drilling jars, bumper subs, etc. The primary purpose of the BHA is to provide weight for the core bit. Other functions

may include providing heave compensation, stabilization for drilling straight (vertical) hole, jarring capability for feeding the BHA when stuck in the hole, and landing shoulders, latches, etc. required for coring operations.

The Ocean Drilling Program utilizes two basic BHA configurations with many special variations depending on the task at hand. The BHA configuration for a particular hole is determined by such considerations as the drilling objective (i.e. desired depth of penetration), coring system to be used, core orientation requirements, type and size of core bit (which dictates the required weight on bit), anticipated hole conditions, etc. Specialty operations such as deploying re-entry cones, casing strings, and/or drilling ahead also affect the BHA selection.

The principal BHA used in spudding single bit holes for the ODP is commonly referred to as the "APC/XCB BHA". This BHA is compatible with most of the specially developed ODP coring systems. It allows spudding (initiating at the sea floor) a sedimentary hole with the Advanced Piston Corer (APC), switching to the Extended Core Barrel (XCB) as the formation becomes more indurated, and finally allows deployment of the Navi-Drill Core Barrel (NCB) coring system for coring in very indurated or crystalline rock types. This BHA also allows deployment of the ODP Pressure Core Sampler (PCS) when it is scientifically desireable to recover core samples under *in situ* "hydrostatic" pressure.

All of these coring systems are interchangeable at any time within the BHA. The APC/XCB BHA, with all required cross-over subs, typically consists of an XCB core bit, one bit sub with lockable float valve (LFV) or one mechanical bit release sub assembly with LFV, one "smooth bore" outer core barrel, one landing/saver sub, one latch sub, one top sub, one head sub, five 8.25-in. drill collars, one 7.35-in. drill collar, and two stands, at a nominal 93.5 ft (28.5 m) per stand, of 5.50 in. "heavy wall" drill pipe. A typical APC/XCB BHA is approximately 423 ft (129 m) long. For oriented coring operations a "non-magnetic" (austenitic stainless steel material with low magnetic permeability) drill collar is substituted for one of the basic alloy steel (AISI 4345) 8.25 in drill collars.

The BHA used most often for coring deep (i.e. greater than 1000 m sub bottom penetration) sedi-

mentary holes, crystalline rock, or for coring basement, is called the "RCB BHA". The RCB or Rotary
Core Barrel is the only coring system compatible with
the RCB BHA. Although there are many variations,
this BHA typically consists of a 9.88 in. (25.1 cm) OD
by 2.31 in. (5.87 cm) ID tungsten carbide, roller cone
core bit (one of several available models), one bit sub
or one bit release assembly (either hydraulic or
mechanical) with support bearing and standard float
valve, one "control length" outer core barrel, one top
sub, one head sub, seven 8.25 in. drill collars, one
7.25 in. drill collar, and two stands of 5.50 in. "heavy
wall" drill pipe. Occasionally additional drill collars
are added if more drilling weight on bit (WOB) is
desired. A typical RCB BHA is approximately 479 ft
(146 m) long.

A set of drilling jars is sometimes included in the
BHA to assist in extracting the BHA from the
formation should it become stuck during the coring
operation. The ODP drilling jars are mechanically
actuated and impart an instantaneous upward/down-
ward force or "jarring" action to the bottom hole
assembly. Their location in the BHA varies according to the specific situation.

Stabilizers are occasionally used to help maintain
vertical hole, aid in drilling in-gauge (accurate and
consistent diameter) hole, and also help to prevent
drill string or BHA sticking.

Bumper subs, used routinely by the predecessor
Deep Sea Drilling Project (DSDP) for heave compensation (adjustment of varying drill string length
due to vertical vessel motion), are not used by the
ODP unless there is a specific "space-out" requirement in the BHA. Heave compensation, aboard
ODP's drill ship *Joides Resolution* (SEDCO/BP 471),
is left to the massive 400 ton (363 000 kg) Western
Gear hydraulic heave compensator located below the
travelling block and above the electric top drive
hung in the derrick. This hardware is commonly
referred to in industry as the "traveling equipment".

Several special BHAs or variations of standard
BHAs can be deployed depending on the operational
requirement. These include such variations as adding
a specially modified (bored out rotor) 9.50 in. positive displacement mud rotor (for bare rock spudding), adding a double-J running tool (for re-entry
cone deployments), using special 9.50 in. (24.1 cm)
drill collars (for more concentrated BHA weight),
etc.

Sandline and Sinker Bar Assembly

All coring systems used in the Ocean Drilling Program are "wireline retrievable". A wireline system
means that the core barrel may be lowered or
pumped down to the bottom of the drill string and
recovered through the drill string utilizing a wireline
or "sandline" as it is often referred to in industry.
This is a much faster method of coring than the
"conventional" system most often used in industry.
With a conventional coring system the core barrel is
attached to the end of the drill string. Cores are
recovered by tripping (recovering) the drill string
itself rather than by tripping a wireline. Use of this
system for continuous coring in 15 000–20 000 ft
(4573–6098 m) water depths is therefore very slow
and would be totally impractical for typical ODP
continuous coring operations.

The sandline used aboard the *Joides Resolution* is
a 0.5 in. (1.27 cm) diameter, extra improved plow
steel, 3×18 swaged wire rope. This rope has a
minimum tensile breaking strength of thirty thou-
sand pounds and is terminated on the end with a
"rope socket". The rope socket adapts the end of the
wire rope to a threaded connection for attaching
other components of the sandwire assembly.

Several pieces of equipment are made-up (attached) below the rope socket. They make up what is
called the "sinker bar assembly" (Fig. 4). This consists of a wireline swivel assembly (to prevent cable
twist), one 3.50-in. (8.89 cm) diameter by 5.0 ft
(1.52 m) longer sinker bar, one set of mechanical
link jars (for jarring loose potentially stuck core
barrels or for shearing the retrieving tool shear pin),
a quick release assembly (to allow the retrieving tool
to be changed rapidly if necessary), and the core
barrel retrieving tool itself.

Sinker bars are used to provide weight to the end
of the wire rope to prevent "floating" while running
in the hole. If the sinker bar assembly starts to float
(fall slower than the sandline speed) the wireline may
be overrun causing the wire rope to become tangled
inside the drill pipe. The sinker bars also provide
weight for jarring.

The core barrel retrieving tool is specially designed
so it will automatically attach to the pulling neck
(top) of the core barrel assembly when it is time to
recover the core barrel from downhole. The ODP
uses two types of core barrel retrieving tools, a "GS"

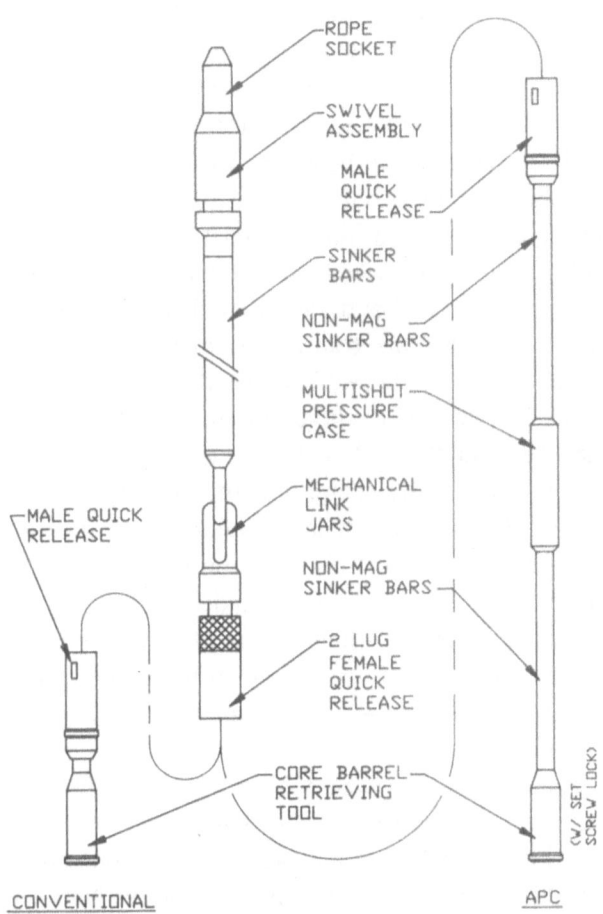

Fig. 4. Typical ODP sinker bar assembly configurations.

deviation and core orientation. The retrieving tool is modified with a set screw to prevent rotation of the multishot pressure case relative to the crew barrel.

Rotary Coring

The rotary core barrel (RCB) coring system is used for routine coring in medium to hard formations including basement and other crystalline rock formations. The RCB recovers a nominal 31 ft (9.5 m) long core, 2.31 in. (5.87 cm) in diameter, and is the most rugged of all the ODP coring systems (Fig. 5). An inner barrel swivel assembly, located at the top of the tool, and a support bearing, located at the bottom of the outer core barrel assembly allow the RCB inner core barrel to remain stationary (non-rotating) during the coring operation. The support bearing also provides a 3.62 in. (9.21 cm) ID landing shoulder where the RCB lands in the BHA and helps to center the inner core barrel with the core bit.

Two inner barrel latch assemblies are available for use with the RCB: a standard (single finger) or a

internal "spear" type or an "RS" external "overshot" type. Each is equipped with a shear pin release system which allows the latching mechanism to be disabled when required. This gives the sandline operator the ability to deactivate the retrieving latch mechanism should the core barrel become stuck in the drill pipe. Without this capability, excessive tension would have to be applied to the wireline to cause it to break. The break would be uncontrolled and therefore could occur at any point in the line, thus possibly causing the loss of a large portion of the sandline and creating significant operational problems for the rig crew while tripping the drill string with the extraneous wire rope inside.

When orienting cores with the APC system two non-magnetic sinker bars with a non-magnetic "multishot" pressure case between them is added into the sinker bar string directly above the core barrel retrieving tool. The pressure case contains a standard industry multi-shot camera used for determining hole inclination,

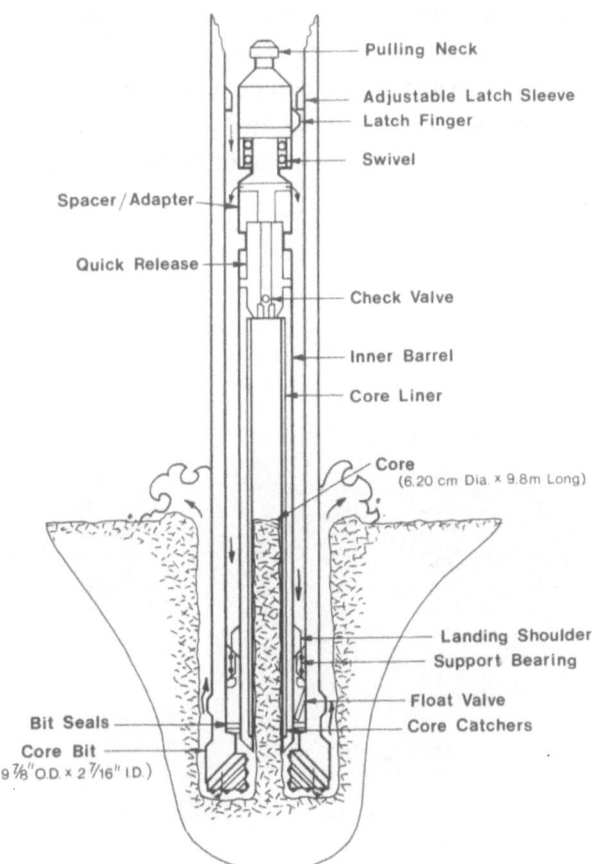

Fig. 5. ODP Rotary Core Barrel (RCB).

special double finger type. Both serve to latch the core barrel down under a latch sleeve located in the outer core barrel assembly, restraining the RCB during the coring operation. The core barrel latch releases mechanically when sandline pull is applied via the core barrel retrieving tool. The bolted latch sleeve, located in the outer core barrel assembly, can be easily adjusted on deck to provide for proper latch down spacing even though assembled tool lengths and outer core barrel lengths may vary.

The RCB is retrieved by the sandline using a standard sinker bar assembly and a core barrel retrieving tool. Two types of retrieving tools are utilized by the ODP: an Otis 3-in. RS-type "external" overshot or an Otis 3-in. GS-type "internal" spear. Both retrieving tools are designed with a shear pin which can be jarred loose to disengage the pulling tool from a stuck core barrel. Mechanical link jars are normally included in the sinker bar assembly to provide additional jarring force for this purpose. The RCB is deployed by "free falling" or pumping it down the drill pipe without the sandline attached.

The inner core barrel contains a clear butyrate core liner. This liner protects the core during the coring operation and also serves as a convenient means for transporting and storing cores once removed from the tool. A core catcher sub, located at the bottom of the inner core barrel, houses one or more core catcher assemblies. Several different types of core catchers can be run depending on the type and characteristics of the formation to be cored (Fig.6).

Washing ahead, or drilling without attempting to recover a core, can be accomplished with a normal RCB barrel in place, called a "wash barrel", or by using a center bit. To make the center bit rotate with the outer barrel the inner barrel swivel is replaced with a drilling sub which locks the upper and lower sections of the tool and allows the drive shoulder on the latch sleeve to transmit torque to the inner barrel. Holes in the wall of the drilling sub allow circulation to the jet holes in the center bit. The check ball in the end of the male quick release is removed when the center bit and drilling sub are used. The double finger latch is considered superior for transmitting the torque to the center bit when rugged drilling conditions are anticipated.

A bit deplugger is often used in place of the center bit when attempting to remove coring debris which occasionally becomes jammed in the throat of the core bit.

Typical core bits used with the RCB are 9.88 in. (25.1 cm) OD by 2.31 in. (5.87 cm) ID tungsten carbide insert (TCI), journal bearing, four cone roller type. The tungsten carbide inserts may be buttons, short chisels, long chisels or a combination of each (Fig. 7).

Fig. 6. Selected Core Catchers—dog, collet and spring-type.

Fig. 7. RCB 9-7/8 TCI roller cone core bit.

Hydraulic Piston Coring

The technique of hydraulic piston coring is based on conventional oceanographic piston coring. A cylinder or tube is pushed into the formation, without rotation, at a high rate of speed. At the same time a piston inside the tube displaces the water in the core barrel out of the top to prevent core erosion and disturbance during core entry. The piston is designed to minimize the back pressure applied to the core during the coring process.

To allow the piston core barrel access to the formation the cones on a standard RCB core bit are spread apart. The resultant bit, with an OD of 11.44 in. (29.06 cm) and an ID of 3.88 in. (9.86 cm), is referred to as "an APC/XCB bit" (Fig. 8).

The primary difference between hydraulic and conventional piston coring is the type of energy source and the magnitude of the applied force. With conventional piston coring a weight is tripped at or slightly above the mudline which falls through the water column, pushing the core barrel into the formation ahead of it. The falling weight provides the energy or coring force. This system is limited to relatively soft formations and shallow penetrations into the seabed (commonly 10–50 m). With hydraulic piston coring the core barrel is landed and sealed off inside the BHA. Rig circulating pumps are used to pressurize the drill string until the preset shear pins fail and allow the core barrel to be

injected into the formation. The stored energy in the drill string provides the dominant source of energy for the coring system. The HPC system generates a large amount of thrust over a very short period of time. With the core barrel accelerating at 6–12 m/sec, the entire coring cycle typically lasts less than two seconds. The benefits of this type of coring system are many. The high rate of speed generated acts to decouple the coring system from the heave induced vessel motion hence there is minimal core disturbance (Fig. 9). The high coring force results in the ability to penetrate into semi-indurated formations at which point rotary coring can be initiated without inducing significant core disturbance. Continued penetration into the formation is facilitated by "washing down" with the primary APC/XCB bit to the point where the previous core run ended its stroke.

The advanced piston corer or "APC" (Fig. 10) utilizes the technology of past DSDP hydraulic (HPC) and variable length hydraulic (VLHPC) piston corers. It incorporates a much more advanced and simpler seal system which results in a 76% greater coring force of 28 000 lb (125 000 N (newtons)). The APC uses a dynamic seal acting between the scoping piston corer and a special honed-bore outer core barrel, commonly referred to as a "seal bore drill collar". The inside diameter of this outer core barrel is 3.80 in. (9.65 cm) minimum, which constitutes the tightest restriction in the BHA.

The non-scoping section of the APC incorporates an adjustable flow-by landing shoulder sub where speed control set-screws can be added or removed to control velocity. An anti-spiral system prevents rotation of the scoping section of the corer relative to the piston rod. An anti-spiral key located in the male quick release tracks down a special groove machined in each of the piston rod sections. All rod sections are interchangeable, even if fabricated separately. When assembled and locked a complete set of piston rod sections, upper, center and lower, will automatically have an aligned anti-spiral groove running the entire length of the piston rod assembly.

The APC recovers core inside a plastic acetate butyrate core liner just as the other ODP coring systems. The core is retained in the barrel primarily with a flapper-type, full-closure core catcher (Fig. 11), although the other core catchers used in the rotary coring system are compatible and can be used if the formation warrants.

Fig. 8. APC/XCB 11-7/16 TCI roller cone bit.

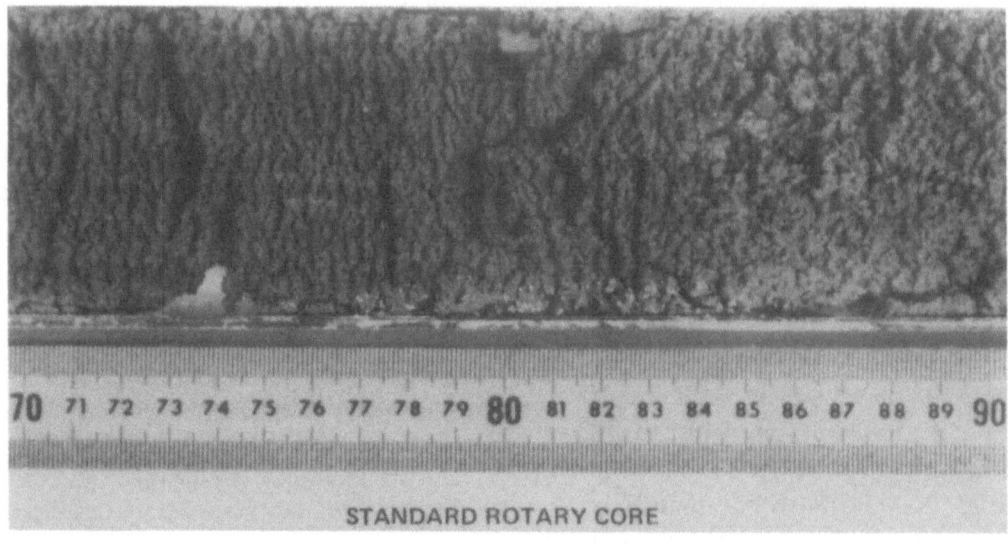

STANDARD ROTARY CORE

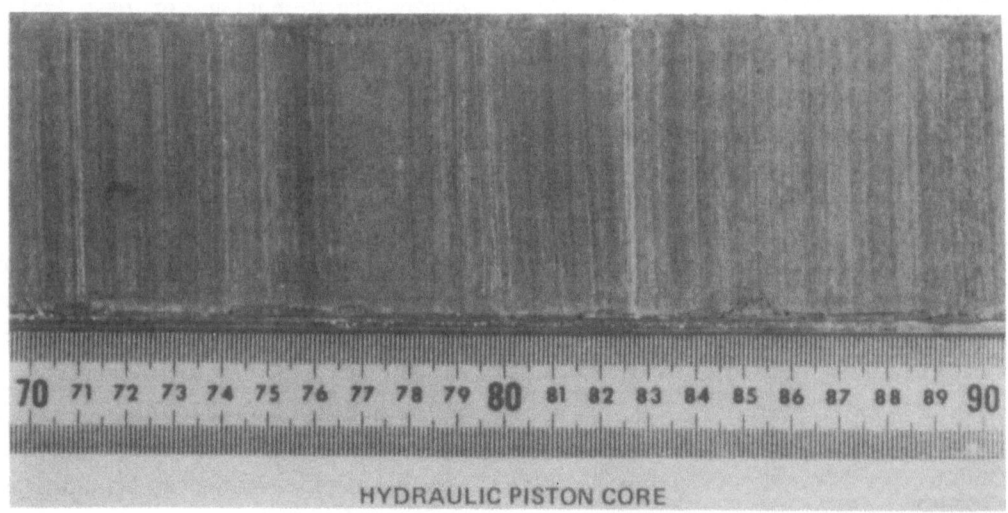

HYDRAULIC PISTON CORE

Fig. 9. Core comparison—rotary *vs.* hydraulic.

A magnetic orientation system is used to determine the orientation of cores taken by the APC with respect to magnetic north. The system consists of an Eastman magnetic multishot survey instrument, a non-magnetic sinker bar system, and a non-magnetic drill collar. Adjustment for baseline alignment between the Eastman multishot and the double reference line on the core liner is accomplished by following two special alignment sequences during the initial assembly of the coring system. Ten to fifteen minutes of additional handling time in the coring cycle are required when orientation data is required.

If desired, an *in situ* temperature measurement can also be taken as part of the routine coring cycle (Fig. 12). A small microprocessor and thermistor can be installed into a special APC cutting shoe and used to record temperature data after the piston corer has been injected into the formation. Thermistor equilibration typically requires waiting 5–8 min before retrieving the core barrel from the formation.

Extended Coring

The Extended Core Barrel (XCB) coring system (Fig. 13) has been developed from a conventional oilfield concept. The XCB is a rotary core barrel

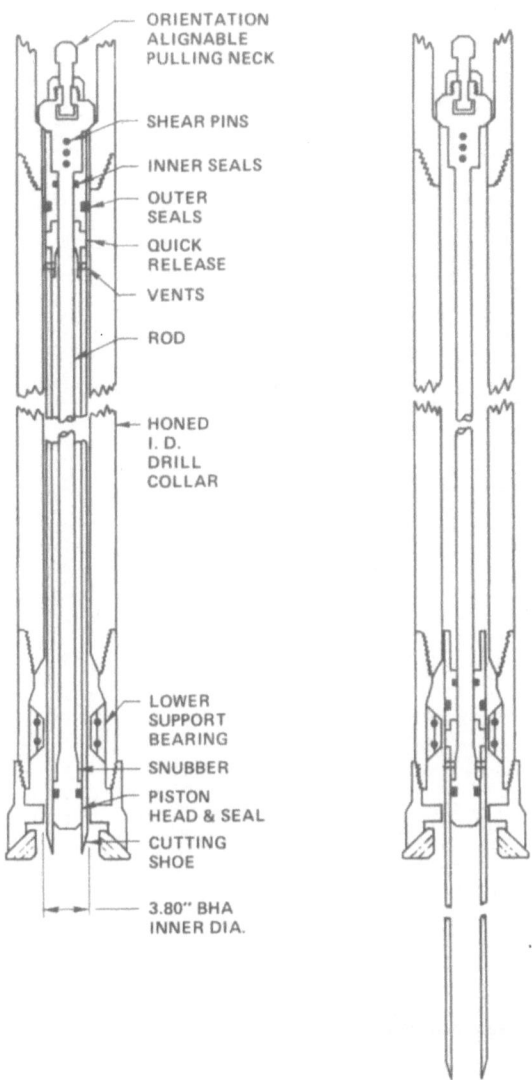

ORIENTATION
ALIGNABLE
PULLING NECK

SHEAR PINS

INNER SEALS

OUTER
SEALS

QUICK
RELEASE

VENTS

ROD

HONED
I. D.
DRILL
COLLAR

LOWER
SUPPORT
BEARING

SNUBBER

PISTON
HEAD & SEAL

CUTTING
SHOE

3.80" BHA
INNER DIA.

Fig. 10. ODP Advanced Piston Corer (APC).

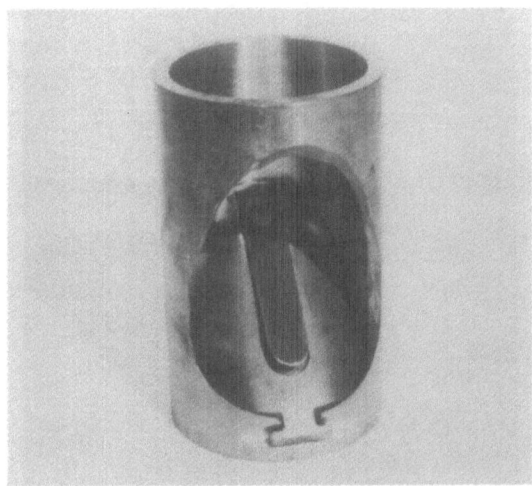

Fig. 11. APC flapper-style core catcher.

system similar in many ways to the standard RCB system. It uses the same butyrate core liner and the core barrel is deployed by free-falling and/or pumping down the drill string. The barrel latches in place in the outer core barrel and accepts a 2.31 in. (5.87 cm) diameter core up to 32 ft (9.8 m) in length. The latch mechanism causes the XCB inner core barrel to rotate with the outer barrel (BHA).

The XCB offers several important advantages over the RCB system. The most significant is its ability to be deployed in the same bottom hole assembly as the Advanced Piston Corer (APC) after piston coring operations have terminated. Rotary coring can therefore be continued without a pipe trip to change over to a different bit and BHA configuration. The bit

used for APC/XCB coring has a 3.80 in. (9.65 cm) throat which is large enough to allow the piston corer to be extended beyond the bit face and into virgin formation. The XCB core barrel has a cutting shoe structure which rotates in conjunction with the roller cone bit and trims the core down to 2.31 in. (5.87 cm). The abrasive drag style cutting action produces better core quality than a roller cone bit in most types of formations.

With the XCB system the cutting shoe can extend about 7 in. (9.83 cm) beyond the face of the core bit. The lower portion of the core barrel is spring loaded over a hex spline to retract into the core bit under a 1500–2000 lb (6700–8900 N) load. In the extended position the cutting shoe shields the incoming core from disturbance caused by the primary bit jet hydraulics. A variable amount of circulation is diverted to the cutting shoe to lubricate, cool and flush the core trimmer. The spring loaded extension/retraction capability provides overload protection of the core barrel assembly and is also helpful when attempting to recover cores from hard and soft interbedded lithologies which cannot be successfully cored with either the RCB or APC systems.

Different types of cutting shoes are available, including a hard-faced "soft formation" shoe and several diamond-type shoes using impregnated, geoset or natural diamond-cutting structures. Combination diamond/saw-tooth bit profiles are also available.

A special XCB latch has been designed for ruggedness, simplicity and positive transmission of torque.

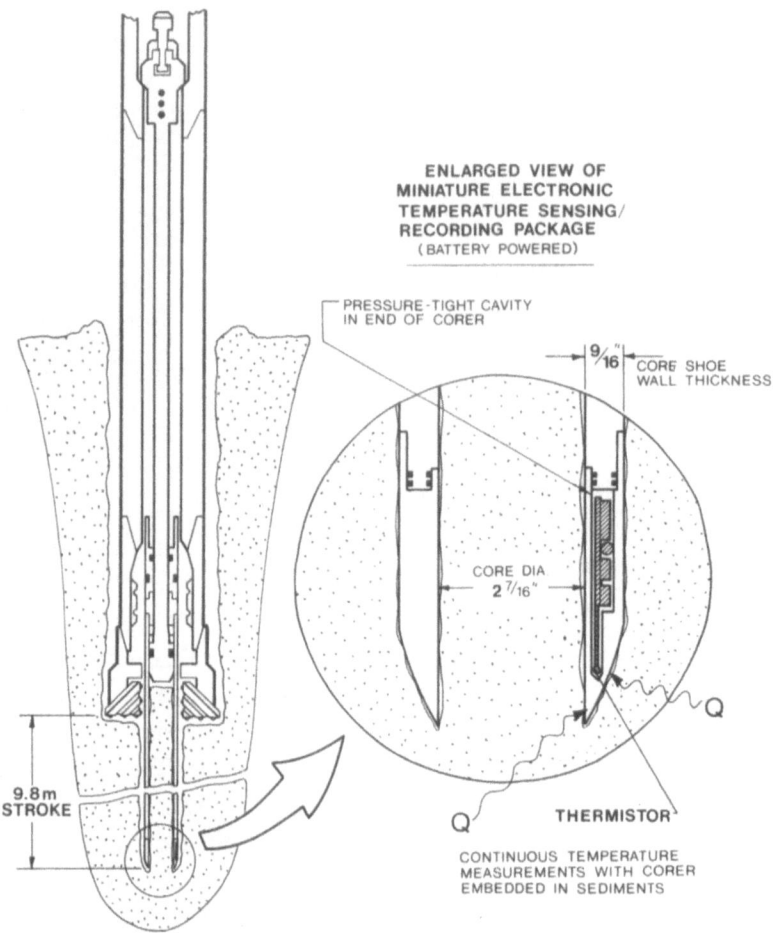

ENLARGED VIEW OF
MINIATURE ELECTRONIC
TEMPERATURE SENSING/
RECORDING PACKAGE
(BATTERY POWERED)

PRESSURE-TIGHT CAVITY
IN END OF CORER

9/16" CORE SHOE
WALL THICKNESS

CORE DIA
2 7/16"

Q

9.8 m
STROKE

Q

THERMISTOR

CONTINUOUS TEMPERATURE
MEASUREMENTS WITH CORER
EMBEDDED IN SEDIMENTS

Fig. 12. APC *in situ* temperature measurement system.

The XCB lands on the 3.0 in. (9.91 cm) ID landing/saver sub, just as the APC. The latch dogs engage in a double-window latch sleeve which ensures that the latch will lock down in any orientation without requiring a "muleshoe" feature.

Major components of the XCB include the latch, hex splined scoping section, spring shaft with helical compression spring, quick release, vent sub, core barrel with non-rotating core liner, and cutting shoe.

The latch was designed and developed specifically for use with the XCB but may be used with any other tool which requires a hold down or torsional transmission mechanism. The latch body is 3.75 in. (9.53 cm) in diameter. The two latch dogs lock out to a 5.0 in. (12.7 cm) diameter. When the XCB enters the pipe the latch dogs are forced up against the spring until they fall into the pulling neck detents. The dogs remain depressed in the 4.12 in. (10.48 cm) bore as the tool travels down the pipe. The pipe bore

widens out to a 5.50 in. (13.97 cm) ID below the latch sleeve in the BHA. When the XCB lands the latch is positioned in this section. The spring forces the dogs to lock out over the high points on the pulling neck. At this point the tool is locked under the latch sleeve and the dogs cannot shift if an upward force is applied to the tool (downward force on top of dogs). When it is desired to release the tool and recover the core barrel, a core barrel retrieving tool is lowered by sandline and locks onto the head of the pulling neck. A pull on the sandline shifts the pulling neck out from under the locking dogs and allows the dogs to fall in and release from under the latch sleeve.

The hex shaped shaft of the male drive sub engages a similar profile in the hex landing sub to provide for up to 8.0 in. (20.32 cm) of axial displacement while continually transmitting torque.

A 34.00 in. (86.36 cm) long helical compression

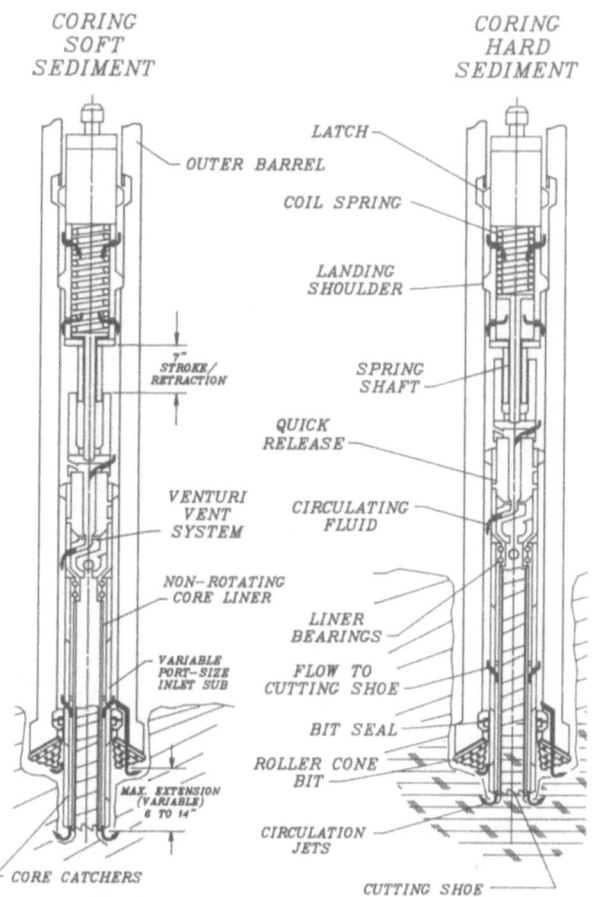

Fig. 13. ODP Extended Core Barrel (XCB).

under evaluation, uses jetting geometry to reduce back pressure at the top of the inner core barrel.

Up to 9.8 m (32.1 ft) of core can be recovered in the standard butyrate core liner. The lower end of the vent sub provides the inner race for a bearing device called the "liner hanger", which is the upper support for the core liner. The liner is also supported at the bottom where it rides on a low-friction bushing.

Several types of cutting shoes are available for use depending upon the nature of the sediment or rock to be cored. The soft formation cutting shoe employs a serrated cutting profile hardfaced with tungsten carbide grit. A portion of the circulation flow to the core bit is diverted to directly lubricate the extended cutting shoe. The flow enters through inlet holes at the top of the shoe, and is directed through an annulus created by the isolation sleeve and out small jet holes at the bottom of the shoe.

spring with a spring rate of 250 lb/in (44 000 N/m) initially maintains the XCB at full extension. A compression force of 2000 lb (8900 N) will cause the spring to fully compress and allow the male drive sub to shoulder on the hex landing sub.

A three-lug quick release is currently used on both the XCB and APC. This mechanism reduces the turn around time between successive cores by providing a rapid means of connecting and disconnecting the core barrel from the upper section of the tool. The male and female sections engage and rotate 50° to lock together.

The vent sub is fitted with a one way check valve to allow fluid to exhaust from the core barrel into the drill string annulus as the core enters. This valve also prevents flow in the opposite direction, protecting the core from being washed out during retrieval. A second pair of slanted holes allows circulation to exit at the vent sub after passing through the quick release mechanism.

An optional "venturi" vent (Fig. 14), currently

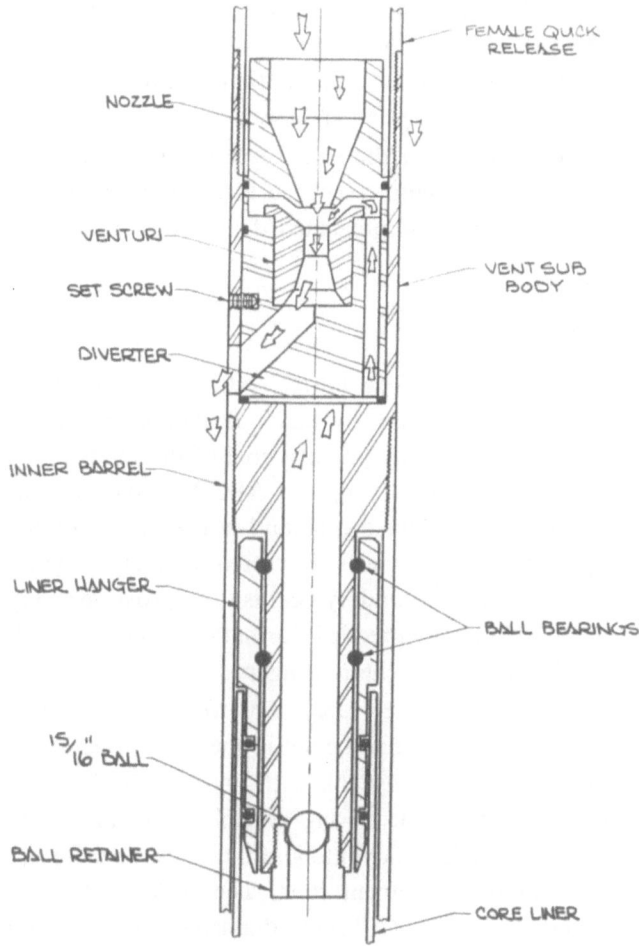

Fig. 14. XCB Venturi Vent Assembly.

Fig. 15. Selected XCB cutting shoes (bits).

The diamond cutting shoes (Fig. 15) may consist of impregnated, surface set, geoset or natural diamonds. Various matrix designs are used. Circulation flow to the cutting edge is similarly diverted down an internal annulus, and may exit directly onto the core or at the face of the diamond bit.

Specially designed combination cutting shoes employ various types of diamonds set in a saw-tooth pattern similar to the soft formation shoe. Yet another experimental shoe uses geoset diamonds imbedded in an impregnated diamond matrix.

Navi-Drill Coring

The Navi-Drill Core Barrel (NCB) is a prototype coring system currently under development by the ODP (Fig. 16). The primary goal of the NCB development is to allow single-bit APC/XCB holes to be extended to greater depths and into more indurated formations particularly fractured crystalline basement rocks. A secondary goal is to improve recovery in hard/soft interbedded formations such as soft chalks laced with chert stringers.

This wireline-retrievable hard rock coring system can be deployed at any point in the coring operation. It is fully interchangeable with the APC/XCB BHA, thus following the coring system to be optimized from the mudline down to and into indurated formations and/or basement rock. The NCB recovers a nominal 2.25 in. (5.72 cm) diameter core, 13.25 ft (4.0 m) long in a plastic polycarbonate liner.

The NCB is comprised of four main components

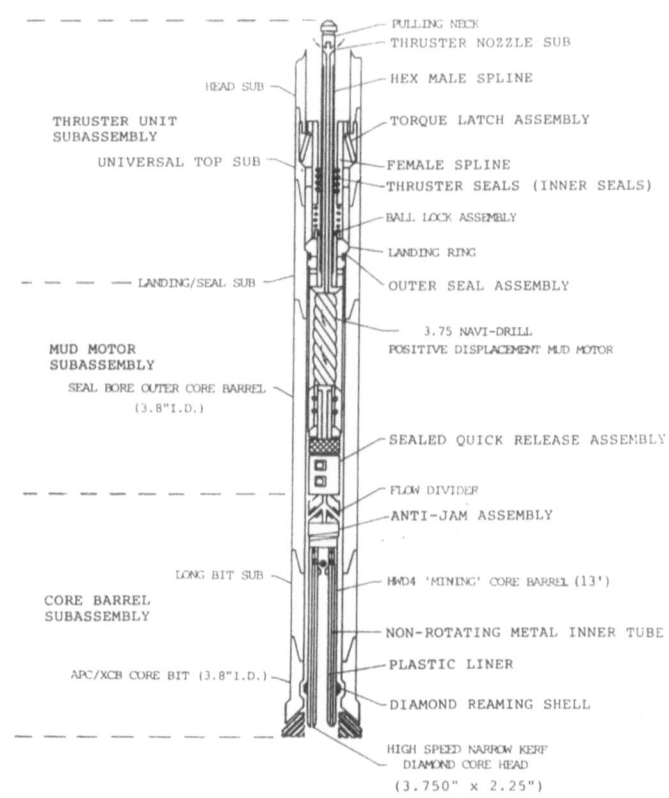

Fig. 16. ODP Navi-Drill Core Barrel (NCB).

or subassemblies: a thruster unit for hydraulically applying weight on bit (WOB), a small 3.75-in. (9.53 cm) OD positive-displacement mud motor for generating downhole rotation (torque), a non-rotating core barrel assembly for receiving the core, and a 3.75-in. (9.53 cm) OD narrow kerf diamond core bit to cut the core.

The thruster unit is comprised of several components performing a wide variety of functions. The primary task of the thruster unit is to translate hydraulic force into mechanical weight on bit (WOB). The pressure drop by circulation through a nozzle sub at the top of a hexagonal spline assembly results in a downward force applied to the diamond core bit. Removable nozzles allow optimization of the desired WOB at various flow rates. The reaction torque generated by mudmotor rotation is transferred through the spline assembly and torque segments to the main outer barrel assembly. When the tool is freefall deployed, the thruster unit dampens the landing impact thus preventing premature unlatching and mechanical failure. The tool may also be deployed using a wireline delivery system; however, this requires an additional wireline trip and results in a less efficient coring operation. The thruster unit also seals downhole causing all of the circulating fluid to be channelled through the mudmotor. In addition, it maintains the stroking portion of the tool (hex male spline, core barrel and diamond core bit) in a latched position until after the tool has landed and rotation is initiated.

The NCB is powered by an Eastman Christensen (EC) 3.75 in. (9.50 cm) OD, 7/8 lobe, Mach I positive displacement mud motor. This motor has been recently developed by EC, primarily for industry oil well drainhole drilling applications. At a pump rate of 170 gal/min (635 l/min) and a pressure drop across the motor of 1160 psi (80 bar), this "drainhole" motor is capable of generating 1250 footpounds (1695 N/m) of torque and 410 rpm. The motor operates at 90% plus efficiency and can develop 96 hydraulic horsepower. The standard core barrel assembly used with the NCB is a modified version of a standard Christensen Mining Products (CMP) HWD4 "HQ" type core barrel. It is attached to the mud motor with a modified (sealed) three-lug quick disconnect to allow handling in the same efficient manner as the other primary ODP coring systems. The barrel contains a non-rotating inner tube which may be run with or without a polycarbonate liner. The core is retained by using either a standard core spring installed in an inner tube shoe (for hard formations) or a special spring loaded dog-type core catcher (for soft formations). Installed between the inner-tube and the core catcher shoe is a breakoff sub which allows for easy retrieval of the core liner. A CMP anti-jam system can be installed directly above the inner tube. In theory, when a core blockage occurs, the inner tube lifts up, energizing the anti-jam system. The resultant "jarring" action is designed to free the blockage and allow unrestricted core entry to resume. Installed directly above the core barrel assembly is a flow divider sub. This sub allows the proper amount of circulating fluid to be directed to the diamond core bit (typically 10–15 gal/min or 38–57 l/min) while the remainder is diverted to the annulus for hole cleaning and cuttings removal. Removable nozzles allow the flow distribution to be optimized as required.

The NCB is designed to operate with several types of narrow kerf diamond core bits (Fig. 17). All bits have an OD of 3.75 in. (9.50 cm) and cut a 2.25 in. (5.71 cm) core. Both hard and soft matrix impreg-

Fig. 17. Selected NCB cutting shoes (bits).

nated diamond bits are available as well as surface set and geoset diamond bits. Appropriate bit selection is determined by the type of formation to be cored. A surface set diamond reaming shell is installed directly behind the core bit to enhance stabilization and help maintain hole gage.

Coring operations commence by freefall deploying the NCB assembly to land within the APC/XCB BHA. As circulation with the rig mud pumps is established rotation of the coring assembly begins inside the BHA (Fig. 18). Increased flow rate causes the NCB coring assembly to release and drop into contact with the formation to be cored (Fig. 19). Coring proceeds until the NCB reaches full stroke at which time the driller sees a pressure spike due to a choking of the flow area within the tool. Coring is suspended and the NCB is retrieved. Additional penetration is made by drilling out the NCB cored "pilot hole" with the main XCB bit and then deploying the NCB system again.

Pressure Coring

The Pressure Core Sampler (PCS) is an ODP developmental coring system capable of retrieving core samples at near *in situ* hydrostatic pressures (Fig. 20). The PCS utilizes both current conventional oilfield pressure coring technology and technology developed by the Deep Sea Drilling Project (DSDP). The PCS is also completely compatible with the APC/XCB/NCB BHA. The PCS is being developed in response to a scientific mandate for retrieving core

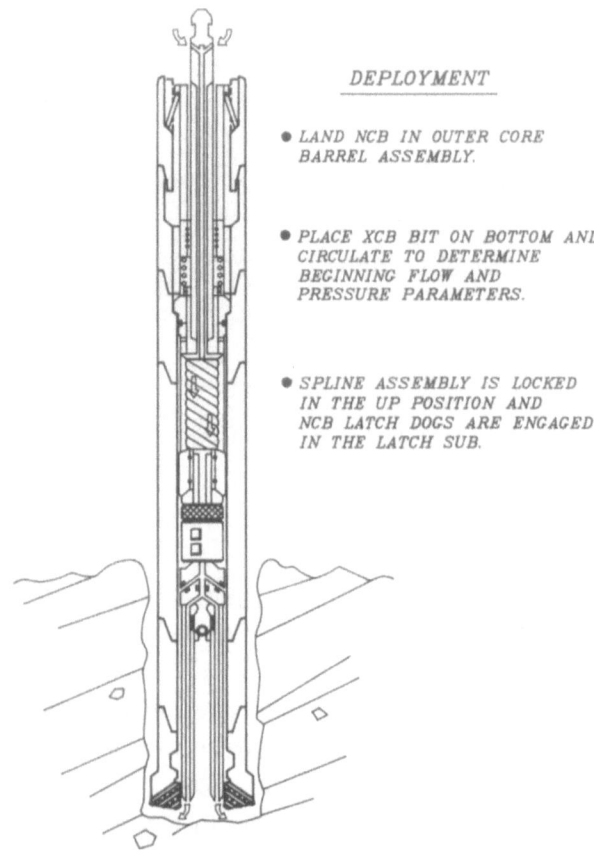

DEPLOYMENT

- *LAND NCB IN OUTER CORE BARREL ASSEMBLY.*

- *PLACE XCB BIT ON BOTTOM AND CIRCULATE TO DETERMINE BEGINNING FLOW AND PRESSURE PARAMETERS.*

- *SPLINE ASSEMBLY IS LOCKED IN THE UP POSITION AND NCB LATCH DOGS ARE ENGAGED IN THE LATCH SUB.*

Fig. 18. NCB deployment—schematic operational sequence.

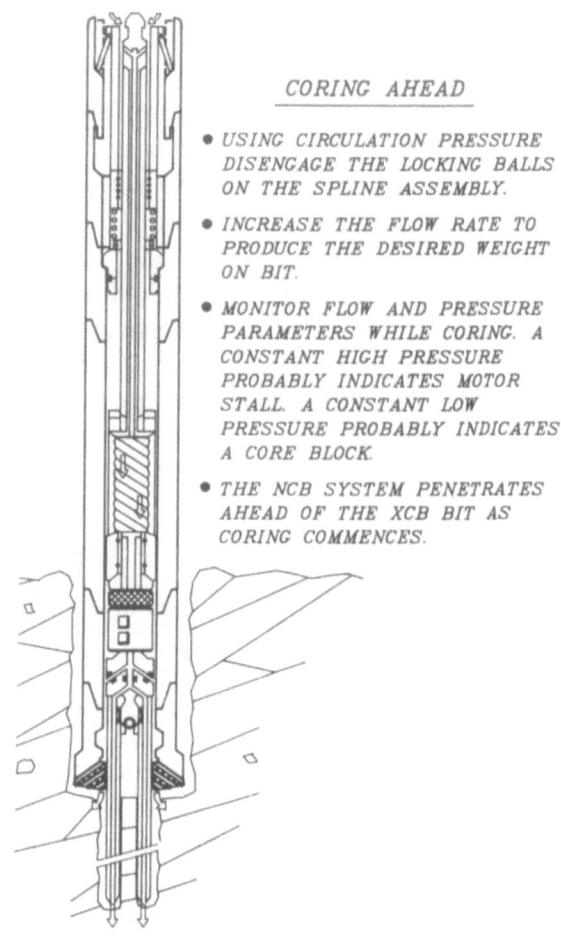

CORING AHEAD

- *USING CIRCULATION PRESSURE DISENGAGE THE LOCKING BALLS ON THE SPLINE ASSEMBLY.*

- *INCREASE THE FLOW RATE TO PRODUCE THE DESIRED WEIGHT ON BIT.*

- *MONITOR FLOW AND PRESSURE PARAMETERS WHILE CORING. A CONSTANT HIGH PRESSURE PROBABLY INDICATES MOTOR STALL. A CONSTANT LOW PRESSURE PROBABLY INDICATES A CORE BLOCK.*

- *THE NCB SYSTEM PENETRATES AHEAD OF THE XCB BIT AS CORING COMMENCES.*

Fig. 19. NCB coring ahead—schematic operational sequence.

samples while maintaining near *in situ* hydrostatic pressures of up to 10 000 psi (690 bar). The system is deployable in soft-to-moderately-indurated sediments and will ultimately have the capability to transfer a 61 in³ (1000 cm³) core sample from the downhole tool to a pressure/temperature controlled laboratory chamber while maintaining downhole pressure. The core sample can then be accessed directly for scientific evaluation under near *in situ* pressure and temperature conditions.

The PCS is a wireline retrievable, free fall deployable, hydraulically actuated pressure coring system. When the PCS is deployed, it lands and latches into the BHA and is rotated with the BHA during coring operations. It is fully interchangeable with the APC and XCB coring systems thus allowing a pressurized core sample to be taken at anytime from the mudline down to indurated formations. The PCS covers a nominal 1.65-in (4.19 cm) diameter core sample, 34 in. (0.86 m) long at pressures up to 10 000 psi (690 atm).

The PCS comprises five main components of subassemblies: latch, actuator, valve-accumulator, ball valve and detachable sample chamber.

The PCS latch subassembly is a modified XCB latch which serves five functions. The latch subassembly contains the landing point for the PCS. The latch subassembly has a 4.0 in. (10.16 cm) outside diameter shoulder which cannot pass the 3.82 in. (9.70 cm) ID throat of the landing saver sub in the BHA, thus preventing the PCS from passing completely through the BHA. By latching into the BHA, the latch subassembly transmits torque from the BHA to the PCS, allowing it to train the core to proper size for entry into the sample chamber. The latch subassembly holds a check ball used in the actuation of the ball valve subassembly. When the latch subassembly is engaged by the wireline and an upward force is applied, it automatically releases a check ball allowing the ball to fall into the actuation subassembly.

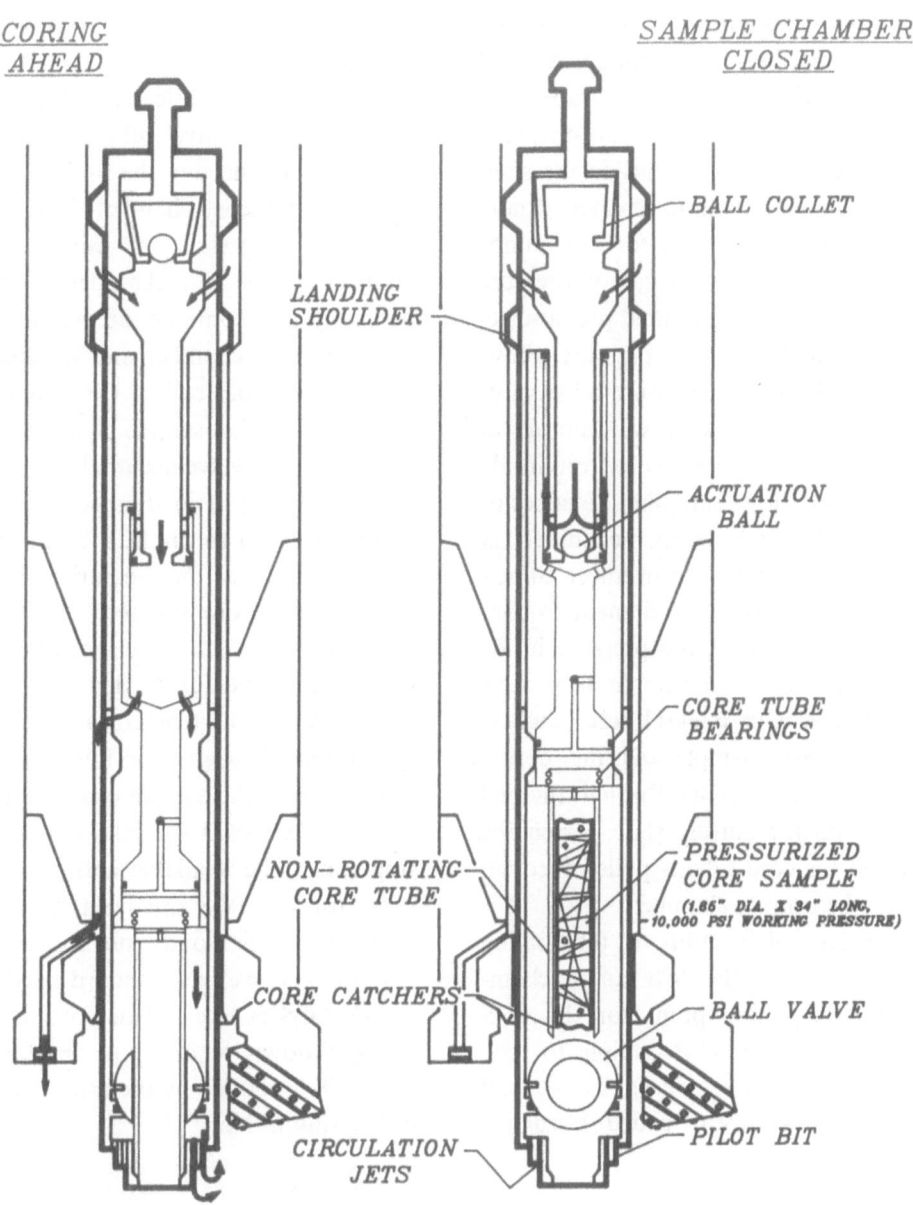

Fig. 20. ODP Pressure Core Sampler (PCS).

Finally, the latch subassembly diverts all flow through the PCS and provides a place for the wireline to automatically attach itself during core barrel retrieving operations. The latch subassembly is attached to the PCS by a three lug quick-release allowing for handling in the same efficient manner as the other primary ODP coring systems. The PCS actuator subassembly serves two functions. It catches the check ball when released by the latch subassembly stopping all flow through the PCS until stroking occurs. Also, when pressure is applied to the PCS and the check ball has been released, the actuation subassembly unlatches and strokes through itself pulling the core tube containing the core sample through the ball valve into the sample chamber. As the core tube is pulled into the sample chamber the ball valve is closed and the upper end of the core tube is pulled into a seal receptacle thus closing the sample chamber at both ends and trapping the core sample at hydrostatic pressure inside the PCS. When the actuation subassembly reaches the end of stroke it latches once again and opens a circulation path through the PCS.

The PCS valve-accumulator subassembly contains a pressure-maintaining mechanism, safety pressure relief mechanisms, a sampling port, temperature and pressure monitoring devices and the core tube. The pressure maintaining mechanism is a built-in accumulator that maintains the pressure inside the sample chamber when a small volume change occurs during sealing. It also compensates for any minor seal leakage should it occur. The safety pressure relief mechanisms include an adjustable pressure relief valve set to automatically vent pressure above 10 000 psi (690 atm). Should the pressure relief valve fail to release pressure a burst disk will rupture at 12 500 psi (862 atm) relieving all pressure from inside the PCS. An access port allows sampling of gasses or fluids directly from the PCS sample chamber. A built-in thermistor and pressure transducer allows for the connection of monitoring equipment to constantly monitor the temperature and pressure inside the PCS sample chamber. The sample tube is a non-rotating metal tube with integral core catchers used to contain the core sample. During coring operations the core tube is extended through the ball valve subassembly into the cutting shoe. When the actuator is activated the core tube is pulled through the ball valve into the sample chamber.

The PCS ball valve subassembly is the sealing mechanism on the bottom of the PCS sample chamber. It also is the connection point for the PCS cutting shoe used to trim the core sample to size. During deployment and coring operations the ball valve is open with the core tube extended through it

into the cutting shoe. When the actuation subassembly is activated and the core tube has been pulled through the ball, the ball is rotated into the closed position sealing the lower end of the sample chamber. The ball valve subassembly also provides a means for connecting the sample chamber to a pressurized laboratory chamber. This is done by removing the cutting shoe and using the threaded end to connect to the test chamber. The ball valve subassembly also contains the pressure containing body of the sample chamber and the seal receptacle used to seal the upper end of the sample chamber.

The detachable sample chamber is made up of the ball valve and valve-accumulator subassemblies. It is 3.75 in. (9.53 cm) in diameter, 5.0 ft (1.52 m) long and is attached to the PCS by quick release connections which allow the pressurized sample chamber to be removed from the rest of the PCS for earlier handling. Since the valve-accumulator subassembly is an integral part of the detachable sample chamber, the pressure and temperature can be continuously monitored. Also, gas and fluid samples can be taken directly from the sample chamber (Fig. 21).

The PCS uses a specially designed pilot-type cutting shoe. The available cutting shoe cutting structures for the PCS are both hard- and soft-matrix impregnated diamonds, surface-set diamonds, geoset diamonds as well as standard hard facing.

The PCS is free fall deployable and therefore is dropped down the drill pipe and landed in the BHA. The PCS is rotated by the top drive via the latch and drill string/BHA.

Fig. 21. PCS tool with sampling manifold.

During coring operations the rig pumps maintain flow down the drill string to keep the hole open and to cool/lubricate the PCS cutting shoe. Once the core has been cut the rig pumps are secured, the wireline is attached to the PCS and an up strain is applied to the PCS latch to release the check ball. The wireline is then slacked off and the rig pumps are restarted slowly, letting the pressure build to activate the actuator and stroke the sample chamber closed. When circulation is once again established the sample chamber has been closed and the PCS is retrieved like any other wireline core barrel. Once on deck the detachable sample chamber is removed from the PCS, placed in a portable temperature controlling bath/safety shroud where temperature and pressure monitoring equipment is attached. The sampler chamber can then be safely moved off the rig floor for scientific evaluation.

Bare Rock Spudding

The development of new technology for spudding holes and coring young, fresh rocks in areas with little or no sediment cover (i.e. mid-ocean ridges, spreading centers, bare rock basins, seamounts, etc.) is a major scientific and engineering objective of the ODP.

The Hard Rock Guide Base (HRB) is designed to sit on the sea floor and provide the necessary lateral support for the bottom hole assembly and bit confinement to spud a hole on bare rock (Figs. 22 and 23). The HRB also serves as a temporary re-entry cone. The guide base is 17 ft (5.18 m) square and 11 ft (3.35 m) tall. The cone inside the guide base is 16 ft (4.88 m) in diameter and 6 ft (1.83 m) deep. Due to the size of the guide base, it is necessary to assemble the HRB in two halves in the ship's moon pool. The HRB is then run through the moon pool vertically on two running cables and once below the keel of the ship, it is rotated into the horizontal running position. The HRB is then lowered to the sea floor suspended from the drill string. Once on the bottom, the HRB and attached cement bags are filled with 2000 ft³ (56.68 m³) of cement to provide additional mass. The HRB is divided by bulkheads into four compartments. The four compartments are isolated from each other so that in the event of HRB damage during deployment the individual compartments insure that at least a portion of the guide base can be filled with cement. The HRB can be deployed in up

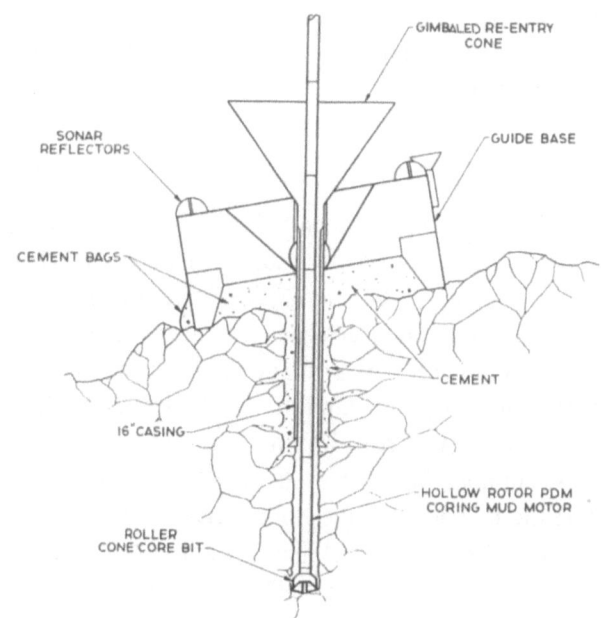

Fig. 22. ODP Hard Rock Guide Base (HRB) schamatic—drilling in young fractured basalt.

to 18 000 ft (5488 m) of water and is designed to land on up to a 20° sloping sea floor with 3 ft (0.92 m) diameter boulders present.

Upon drilling/coring 100–200 ft (30–60 m) of surface hole through the guide base, a special gimbaled reentry cone with a string of 16.0 in. (40.6 cm) or 11.75 in. (29.8 cm) casing is run and cemented in

Fig. 23. HRB photograph.

place. The purpose of the gimbal is to provide adequate surface contact to support the reentry cone at angles of up to 20°.

Mud Motor Coring

The Positive Displacement Coring Motor (PDCM) is a modified version of a standard Eastman Christensen 9.5 in. (24.1 cm) OD-positive displacement mud motor (PDM) designed for use with the hard rock guide base. The primary modification required replacement of the standard "solid rotor" with an "inverted hollow rotor" (Fig. 24).

A nominal 30 ft (9 m) core barrel deployed via wireline is landed in the hollow rotor section of the motor. The core barrel has 3.13 in. (33.4 cm) OD upset connections and a 2.92 in. (7.42 cm) OD mid-

body with a 2.50 in. (6.35 cm) ID. A plastic liner is run inside the core barrel to allow easy removal of the core from the barrel. The rotor is attached to the drill string through the drive subassembly. Torque is generated as fluid mud passes down the drillstring and between the rotor/stator assembly. This torque rotates the 40 ft (12.2 m) long outer motor housing, which in turn induces torque and rotation to a 10.5 in. (26.7 cm) OD by 2.25 in. (5.72 cm) ID roller cone core bit. After cutting a 30 ft (9 m) core, the core barrel is retrieved by wireline, completing the coring cycle.

The PDCM provides up to 6000 foot-pounds (8077 newton-meters) of torque at a flow rate of 600 gal/min (757–2271 l/min). Bit rotational speeds of 90–120 r.p.m. can be generated by the motor with a maximum pressure drop across the motor of 640 lb/in.² (44 bar).

With this system the entire drill string remains stationary. Only the core bit and the motor housing rotate. The enhanced stability of the drill string and BHA as a result of using the coring motor allows exploratory coring operations to be performed on bare rock efficiently and reliably.

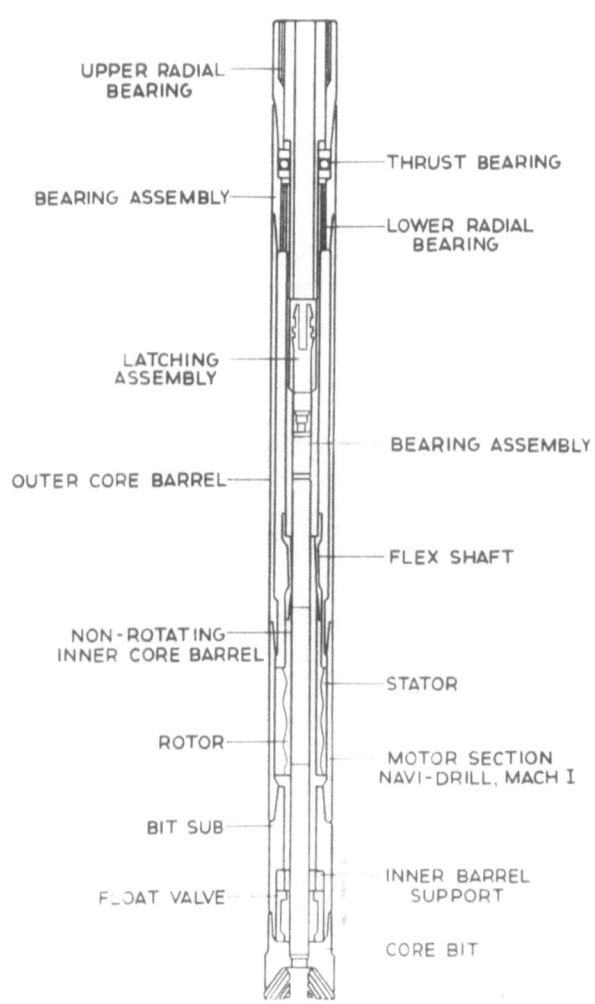

Fig. 24. Positive Displacement Coring Motor (PDCM).

Fig. 25. Diamond Coring System (DCS) top drive concept.

Diamond Coring

The diamond coring system (DCS) is a developmental coring system designed to drill and core both sedimentary and crystalline rock formations. A scaled-down version of the DCS was tested during January 1989 in the Luzon Strait, just north of the Philippines. The purpose of the test was to evaluate the potential and validate the use of a top-driven high-speed diamond coring system deployed from a floating vessel.

This unique coring system involves running a small-diameter drill rod string inside 5.50 in. (13.97 cm) OD drill pipe to the sea floor (Fig. 25). This "working" string has a 3.50 in. (8.89 cm) OD

pipe body and a 2.94 in. (7.47 cm) ID with 3.87 in. (9.83 cm) OD upset connections. A 4.0 in. (10.16 cm) OD by 2.40 in. (6.10 cm) ID high-speed diamond coring bit is run with a core barrel assembly on the drill rod string. The drill rod string is presently rotated with a hydraulic top drive at 60–500 r.p.m. An electric top drive is currently under evaluation as a successor to the hydraulic unit. The small-diameter, narrow kerf, diamond bits are typically operated with 2000–12 000 lb (908–5448 kg) of drilling weight.

A secondary heave compensator system is used to maintain precise control of weight on bit. This secondary "active compensator removes load fluctuations resulting from the mechanical inefficiencies of

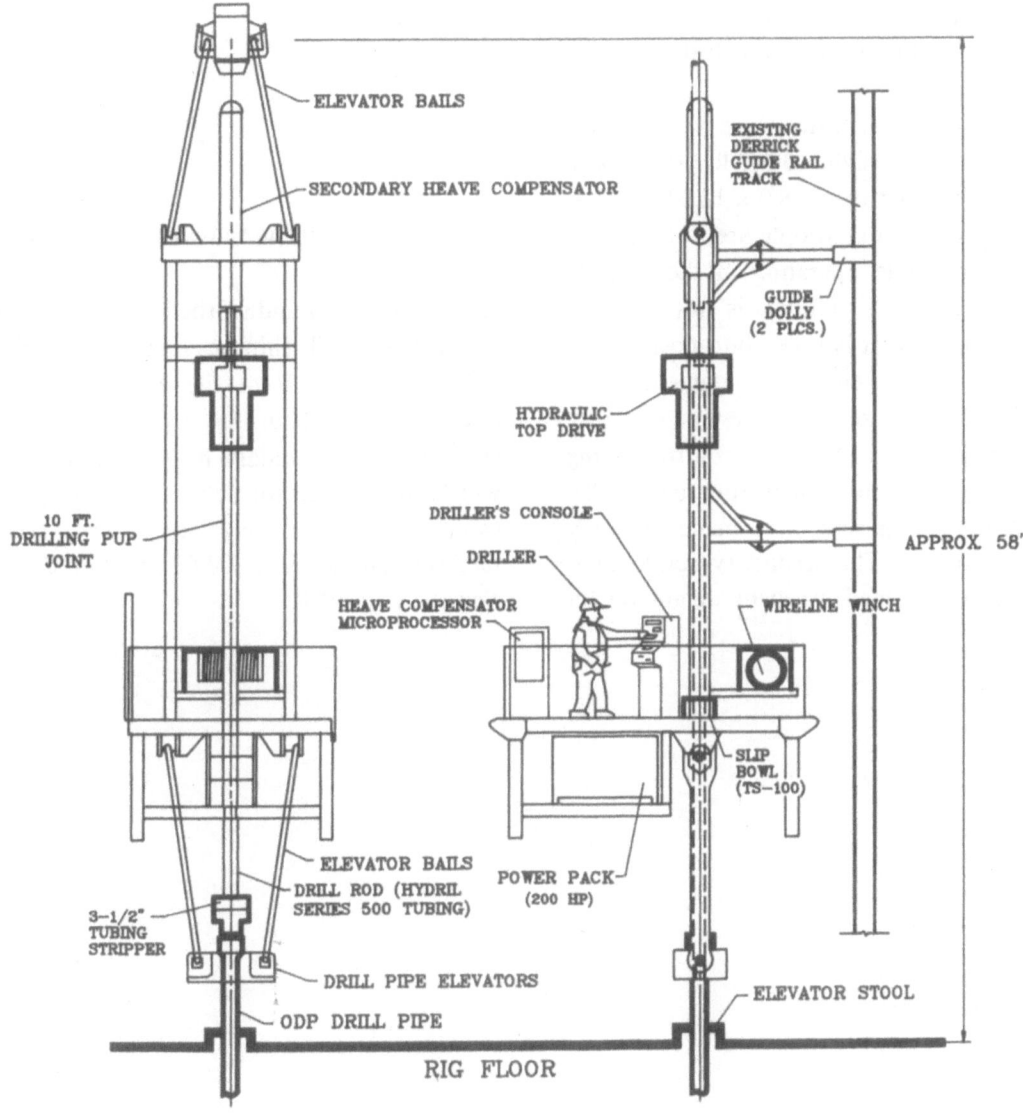

Fig. 26. DCS platform configuration.

the primary 400 ton (363 000 kg) "passive" heave compensator. The result is weight on bit control for the small diamond core bits of ± 500 lb (227 kg).

All of the diamond coring operations and drilling functions are controlled and conducted from a manned platform, suspended in the derrick (Fig. 26). From the driller's console on the coring platform, the driller operates the top drive, secondary heave compensator, wireline winch and controls the make-up/break-out of the drill rod joints. Upon tripping the drill rod joints to bottom, the driller activates the secondary heave compensator and automatic feed system. At that point, the diamond core bit is automatically fed to bottom and the desired bit weight is established for the coring run. Upon completing the coring run, the bit is retracted off bottom and the core barrel is retrieved. A empty core barrel is dropped (free-fall deployed) down the drill rod string and coring is resumed.

The "manned" diamond coring system platform is 45 ft (12.43 m) tall and weighs 40 000 lb (18 160 kg). The work area on the platform is 8.0 × 12.0 ft (26.2 × 39.4 m) square. Two-to-four people are stationed on the DCS platform while operating in the derrick. When not in use the DCS platform is stored out of the way on the rig floor. It is rolled into position via a portable dolly/track system for deployment.

A scaled-down diamond coring system was successfully deployed with a 1700-m drill rod string during ODP leg 124E in January/February 1989 (Fig. 27). Several coring runs were made in heavy seas, with good results. The secondary active compensator system maintained excellent weight on bit

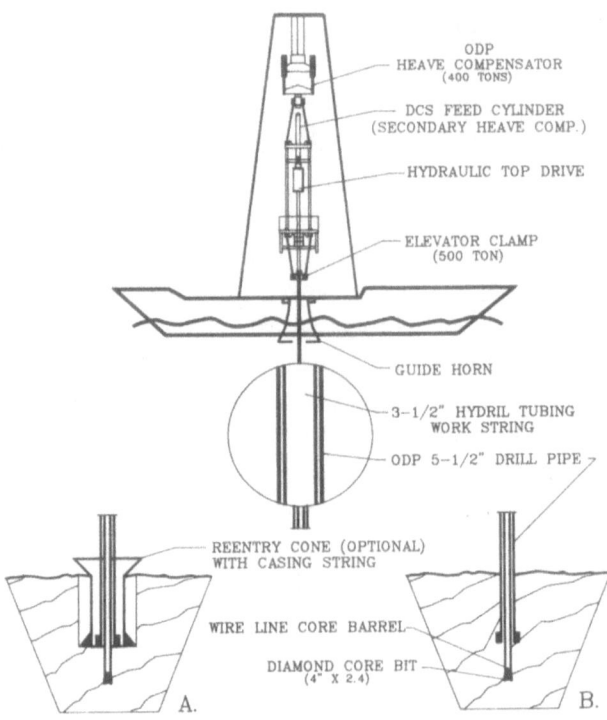

Fig. 27. DCS platform, Varco top drive, and Hydril tubing.

control, allowing undisturbed sedimentary cores to be cut with the drill ship heaving significantly.

During Phase II of the diamond coring system development, the drilling systems (top drive, feed cylinder, and secondary heave compensator system) will be redesigned for deployment in up to 14 760 ft (4500 m) of water. This deep-water system is scheduled for testing on the ODP drill ship *Joides Resolution* early in 1990.

The Status of Geological Dredging Techniques

R. B. KIDD[1], Q. J. HUGGETT[2] and A. T. S. RAMSAY[1]

[1]*University of Wales, College of Cardiff, PO Box 68, Cardiff CF1 3XA, UK*
[2]*Institute of Oceanographic Sciences, Deacon Laboratory. Wormely, Godalming GU8 5UB, UK*

(Received 27 April, 1989; accepted 1 September, 1989)

Key words: geological dredging, basement rock samples, pinger telemetry, transponder navigation, swathe mapping, sorting criteria, dredging and drilling comparisons.

Abstract. Scientific sea-floor dredging is currently used in marine geology primarily by the "hard-rock" community interested in the recovery of basement rock samples from the unsedimented deep ocean floor. The technique has generally been eclipsed by ocean drilling for recovery of sedimentary rocks, because of perceived uncertainties in the location of sampling and in the representativeness of recovered material. This contribution reviews dredging equipment currently in use by marine geological institutions and refers to pinger attachments that allow precise information on the behaviour of the dredge to be telemetered back to the ship. We argue that improvements in ship navigation and transponder navigation at the seafloor, when used in conjunction with surface and/or deeply towed sidescan and swathe-mapping surveys, now allow for considerably less uncertainty on the location of dredge sampling. Refined sorting criteria for dredge hauls are now also available. Recent comparisons of regional sample recovery by ocean drilling and by dredge sampling indicate that the dredge hauls can usefully supplement the drilling data in the construction of sedimentary and tectonic histories of seafloor areas.

Introduction

Marine scientists have, since the earliest days of oceanography, used a variety of towed equipment to sample the deep ocean floor. Dredges were used in the 1860s on the four-year voyage of HMS *Challenger* to obtain biological and geological samples from all the major ocean basins. The basic design of dredges for sampling seaward of the continental shelves have changed little over the intervening century (Fig. 1). A solid metal frame "mouth" is towed on a wire rope from a deep-sea winch on the ship. Behind the frame is a collecting bag, usually of interlinked metal grommets. The dredge is weighted to stay in contact with the sea floor either by trailing weights on a heavy chain behind the bag or with a single lead weight towed ahead of the frame. A range of variations on this basic design are in current use. The main purpose of this chapter is to review techniques in dredge sampling for geological purposes, including operations and navigation at sea, monitoring of the dredge, and assessment of rock recovery.

Sediment coring has become the marine geologist's main sampling tool for detailed analysis of the (generally Late Quaternary) record held in the uppermost tens of meters of unlithified deep sea sediments (Weaver and Schultheiss, this volume). Dredging was, until the late 1960s, the only means available for sampling older lithified sedimentary rocks or the igneous oceanic basement. Great care was taken through accurate navigation, the use of various methods of geophysical surveying and monitoring of dredge behaviour, to ensure that dredge sampling was as representative as possible of the geological terrains being investigated. However, through the 1960s and 1970s, none of these techniques had advanced sufficiently to eliminate uncertainties associated with this sampling method.

The advent of scientific ocean drilling, begun in 1968 by the *Glomar Challenger*, made it possible to sample continuous sedimentary rock sections to basement and ushered in a period when dredging was regarded as a relatively poor tool on which to base studies of the sedimentary or tectonic history of an oceanic area. The Deep Sea Drilling Project (DSDP) was never able to drill directly into igneous rock over the fifteen years of *Glomar Challenger* operations and it is only since 1985 that its successor, the Ocean Drilling Program (ODP), could offer a supported hard-rock spud-in system to accomplish this task (Storms, this volume). Because of this and because the drillship is only infrequently committed to such basement studies, the marine "hard rock" petrology

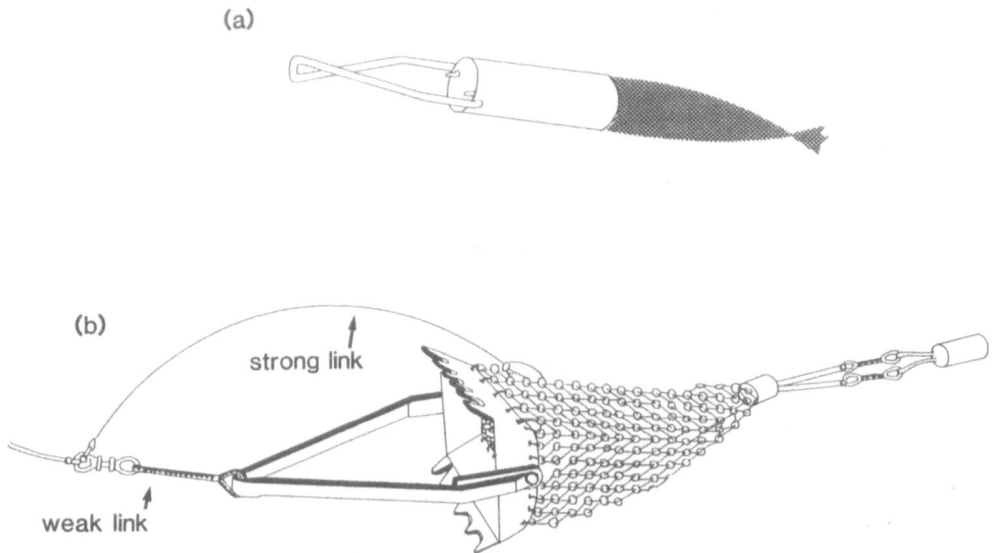

Fig. 1. Two commonly used dredges: (a) Pipe Dredge. This can either be open ended with a collecting bag or closed for use over recent lavas (see text). The mouth diameter is usually around 50 cm. (b) Rock dredge. Based on the Nalwalk design (Nalwalk *et al.*, 1961), dredges like this form the mainstay of geological dredging operations.

community has retained dredging as its primary sampling tool throughout the past two decades, supplementing it more recently with submersible operations (e.g. Fox and Stroup, 1981). Drilling operations, also, cannot satisfy the need for large-volume aerially distributed samples for use in investigations of regional diversity.

Over this same period, techniques of swath mapping (Grant and Schreiber, this volume); accurate navigation and near-bottom survey have advanced markedly. Furthermore, results are now available from precisely-positioned continuously-drilled sections in areas where knowledge of the geological history previously had been based only on dredge surveys. This enables us to consider the representativeness of dredge haul material in regional geological studies. Examples are given later in this review,

together with an assessment of the status of geological dredging in the late 1980s in the light of the developments in drilling and submersible technology.

Types of Dredges

Early dredges were designed to collect for study whatever material was present on the sea bottom, be it biological or geological. Over the years dredge design has diversified and the more sophisticated dredging equipment is presently used by the biological community (Fig. 2). However, parallel developments in techniques for navigating the dredge and its behaviour have occurred for both applications of dredging. In some cases, biological benthic dredges have been used by geologists, for example to study ice-rafted and other rock debris on abyssal plains

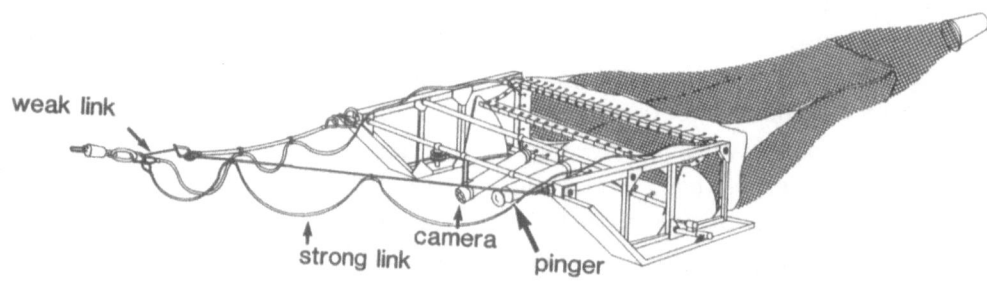

Fig. 2. Benthic sledge used for sampling sediment surfaces. Used mainly by biologists for sampling benthic communities, this dredge will also collect manganese nodules and other rocks lying on sedimented seafloors.

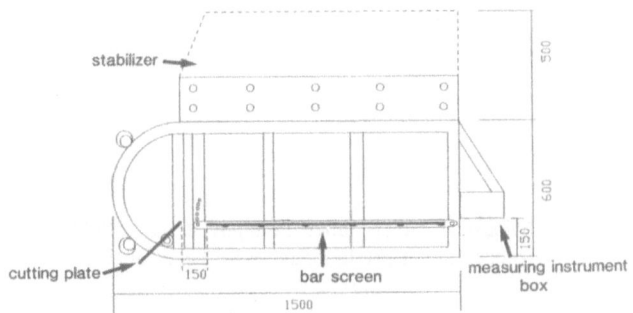

Fig. 3. Prototype sampling sledge specifically designed for manganese nodule collecting, dimensions are in centimeters (after Kinoshita *et al.*, 1975).

(Kidd and Huggett, 1981). Some of the largest dredges aimed at rock collection, with frames over 1.5 m-long, were designed as part of prototype systems (Fig. 3) for manganese nodule mining, again for use on sedimented abyssal plains (Cronan 1980). We shall concentrate here on dredges designed to sample rock outcrop in geological studies of continental margins, mid-ocean ridge systems and other sea floor physiographic features.

Geological dredge design has remained very simple, largely because of the logistical constraints of operation in rugged sea floor terrain and the high risk to sophisticated equipment when the primary aim is to break off and recover *in situ* material for study. Three basic designs have had widespread use:

(a) pipe dredges
(b) rectangular dredges
(c) cylinder dredges

(a) PIPE DREDGES

These relatively small dredges comprise a metal pipe linked to the trawl warp (tow-cable) through a metal bail. The pipe frame is generally around 15 cm in diameter and 45 cm long (Fig. 1a). A net collecting bag completes the dredge, whose total weight is around 25 kg. Although popular for relatively rapid reconnaissance sampling of sea floor features until the early 1960s, pipe dredges are now rarely used.

(b) RECTANGULAR ROCK DREDGES

These are the most commonly used dredges. Most rectangular-frame rock dredges (Fig. 1b) have evolved from, or have similar basic components to that described by Nalwalk *et al.* (1961). The specific aims of the Nalwalk design (Fig. 4b) were: (a) that it

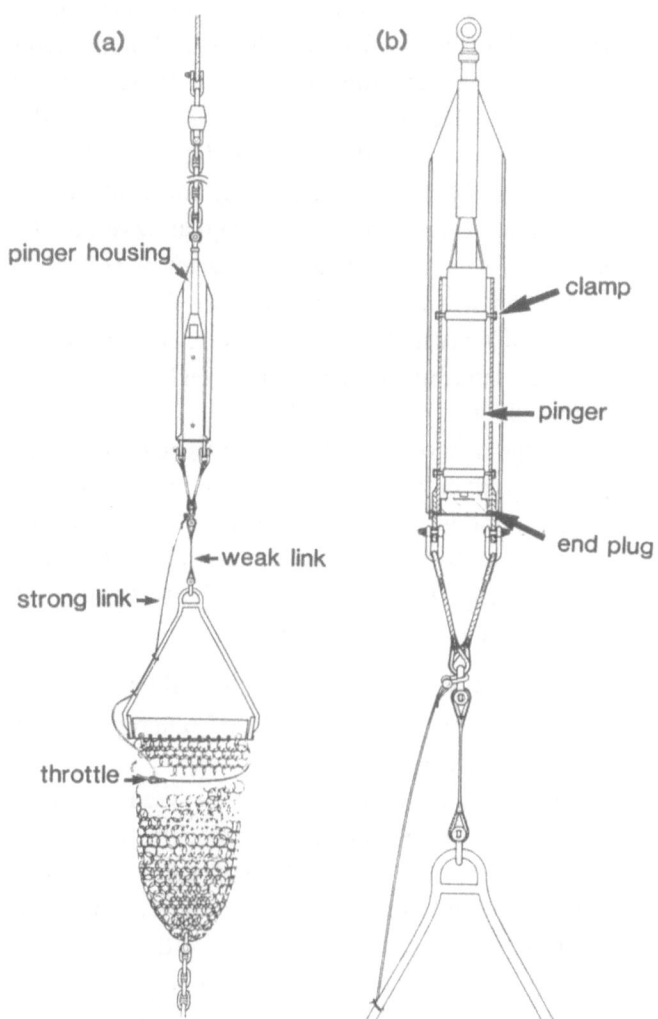

Fig. 4. Dredge rig currently used by the Institute of Oceanographic Sciences, U.K. (after Gaunt and Wilson, 1975): (a) The complete rig with weak and strong links. (b) The pinger housing (cutaway drawing).

should collect large numbers of small cobble-size samples; and (b) that it should be able to free itself from a rocky bottom. The bail linking the dredge frame to the trawl warp was free to swivel and this permits the dredge to ride over large boulders or massive outcrops without becoming snagged. One side of the Nawalk dredge frame or mouth was weighted with lead so that this side dragged the bottom. Also a weight was towed at the rear of the system to stabilize the chain bag in deployment and towing. Weak-links were used in the system to prevent the tail weight "hanging up", if it became caught in the bottom, and to allow the bail to break free if the frame became snagged. The weak-link was of rope, with a swivel on each end and was rigged between the ship's cable and the bail. A light chain was led in parallel from the ships cable end of the wire rope to the chain bag such that a break at the weak-link would overturn the dredge and allow its recovery without dumping the bag and contents. In this design varying amounts of angular swivel at the base were used depending on how much soft sediment cover was expected in the sampling area.

Sizes of rectangular rock dredge vary from 38 cm to 90 cm mouth-width and total weights vary from 30 to 120 kg.

A range of modifications on the basic design have been utilized: cutting teeth have been included on the dredge frame mouth (Fig. 1b) with a view to breaking off outcrop samples. Some systems, including a variation used by the Lamont Doherty Geological Observatory (LDGO), include:

- a lead ball weight placed ahead of the bail of the dredge, to try to break samples off outcrop prior to the dredge passing over them;
- nylon nets of varying mesh size used as inner linings for the metal grommet collecting bag. These control the size of material collected. The mesh linings cannot be too fine or they restrict water flow through the bag and it soon fills with mud;
- heavy anchor chains sometimes used instead of a single tail weight. In some systems two chains are towed in parallel from a box-shaped grommet bag.

Systems of weak-links in series are frequently a feature of rock dredging operations because of the high incidence of dredges becoming stuck on the bottom. These weak-links ensure failure in stages should the dredge "hang up" and put the ship's trawl warp at risk of parting.

The standard rock dredging gear used over the years by the British Institute of Oceanographic Sciences was an example that used weak links and was designed to collect a wider range of sample sizes than the basic Nalwalk design. Between the frame and the towing bridle, a pin that sheared at 1 ton strain allowed the dredge to tilt at 30° (Burt, 1979). A 2.5-ton weak link was inserted between the dredge bail and the chain connection to the warp. A 13-mm wire ran as a noose from this chain through the front part of the grommet bag so that when the 2.5-ton link broke, the bag would be throttled and the frame was turned over from the rear which usually released

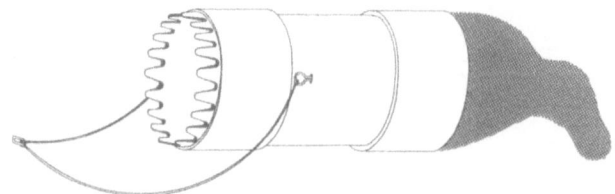

Fig. 5. Cylindrical Dredge after Fillippov et. al. (1970). The dredge is 1 metre in diameter by 1.5 m long.

it. A 4-ton shearing link was the last line of defence, situated at the end of the trawl warp. If the dredge had failed to clear, by sailing the ship on an opposite course or by shearing of the lower strength links, this last link ensured loss of the dredge but kept the warp intact. Figure 4 shows a later version of the IOS system, modified for use with a tilt-switch pinger, but still incorporating some of the weak-linked system.

(c) CYLINDER ROCK DREDGES

Cylinder dredges are most commonly used by Russian research institutions, although they are in use also by American (e.g. the United States Geological Survey Marine Geology Group) and Japanese institutions. The design of the cylinder dredge is again simple but the frame ahead of the collecting bag is an elongate cylinder. The design was an attempt to circumvent perceived problems with early rectangular dredges relating to rigid towing bails, short frames and generally low weight, which together often resulted in these early devices bottoming in a non-operating position or adopting an orientation during towing in which the mouth would not sample throughout the operation. Later rectangular dredges as described above have swivelling bails, weighted lower frames and release mechanisms, but the simple operation of the cylinder dredge is still favoured by some groups, especially in reconnaissance studies.

The cylinder dredge described by Fillippov et al. (1970) is a typical example (Fig. 5). The main component is a 12 mm-thick steel cylinder, 1 m in diameter and 1.5 m long. Two 12 mm-thick rings are welded to the cylinder's outside surface, providing additional strength. Rounded cutting teeth are an integral part of the forward end of the cylinder. Two lugs of round steel rod provide towing attachments, welded to the outside of the cylinder forward of its centre of gravity. A series of drilled holes at the tail end provide securing links for the net collecting bag.

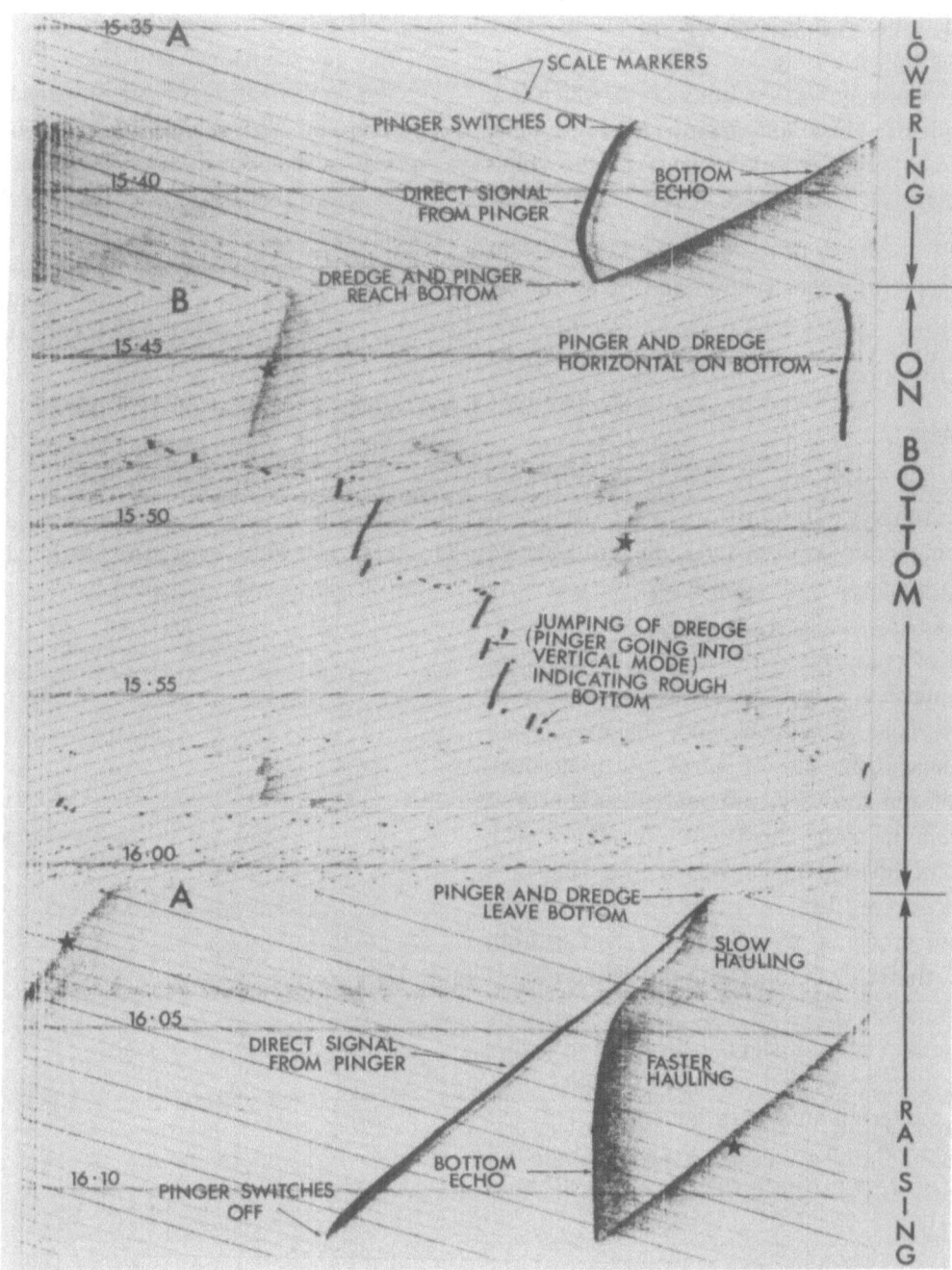

Fig. 6. Pinger record from a dredge station where several "bites" were felt. (After Gaunt and Wilson, 1975).

This conical bag is protected against tearing by an ox-hide cover and narrows to a towpoint for a 50 kg tear-drop tail weight. The dredge is towed simply from chains connecting the side lugs to the ship's cable. The overall weight of the system is 650–750 kg.

The advantages of the simpler cylinder dredge over less sophisticated versions of the rectangular dredge were seen to be:

1. its smaller contact area with the substrate and the relatively large weight applied to this area to engage or scrape rock projections;
2. its shape which ensures that it lands and is kept in an operating position on the sea floor;

3. the flexibility of the towing arrangement which allows it to roll sideways rather than hang up on a projection; and

4. the relatively weak cutting teeth which will bend or break off before tensions build enough to threaten breakage of the ship's cable.

The greatest attributes of the cylinder dredge are its simplicity, its strong construction and its reliability.

Dredging Operations

THE TARGET

Rock dredging operations are carried out in rugged sea floor terrains where slopes are usually sufficient to ensure only a thin cover of soft sediment. Detailed regional bathymetric and seismic surveys will have been run and the dredging is usually aimed at sampling particular sequences identified as outcropping on seismic profiles. Targets are mainly fault scarps or the eroded sides of canyons and channels. Rock debris or talus at the base of scarps can often make up most of the recovery, although freshly-broken material from the scarps themselves provide the ideal sample. Some investigators involved in submersible operations believe that so much of the material recovered in dredging at mid-ocean ridges is from talus piles that its usefulness in interpretation is seriously impaired (e.g. Fox and Stroup, 1981). As will be described later, dredging operations in higher latitudes recover additional ice-rafted material from terraces in the terrain and from talus piles. This has been transported and dropped by melting icebergs. These lithified rocks require careful sorting from *in situ* material.

OPERATIONS IN GENERAL

Once the targets have been selected, ship operations are generally designed to time the dredges' arrival at the bottom such that the ship can then take a course at right angles to the feature and make the dredge haul over the scarp. If at all possible operations are carried out head-to-wind, recognizing that the catenary of the ship's tow cable or trawl warp will put the dredge a kilometer or so behind the vessel.

Typically, around 8000 m of cable is paid out for dredging operations at around 5000 m water depth. However, information on the length of cable out and

the depth below the ship are of little use in trying to calculate true dredge position.

The key elements in any successful dredging operation are navigation of the ship *and* of the dredge system and also monitoring of the behaviour of the dredge at the bottom. It is clearly important not only to recover rocks but also to have information on where in the terrain they were sampled from. Even the crudest operations depend on a pinger placed near the dredge (Fig. 6) to transmit information from depth and a tensiometer to record load on the ship's cable and from this dredge behaviour and likely sampling times from the recorded "bites" (Fig. 7).

Although improvements in navigation and in monitoring dredge behaviour have been spectacular, there still remains a certain mystique to dredging operations. Thus, even in the early days of tensiometer measurements, practitioners would still sit on the

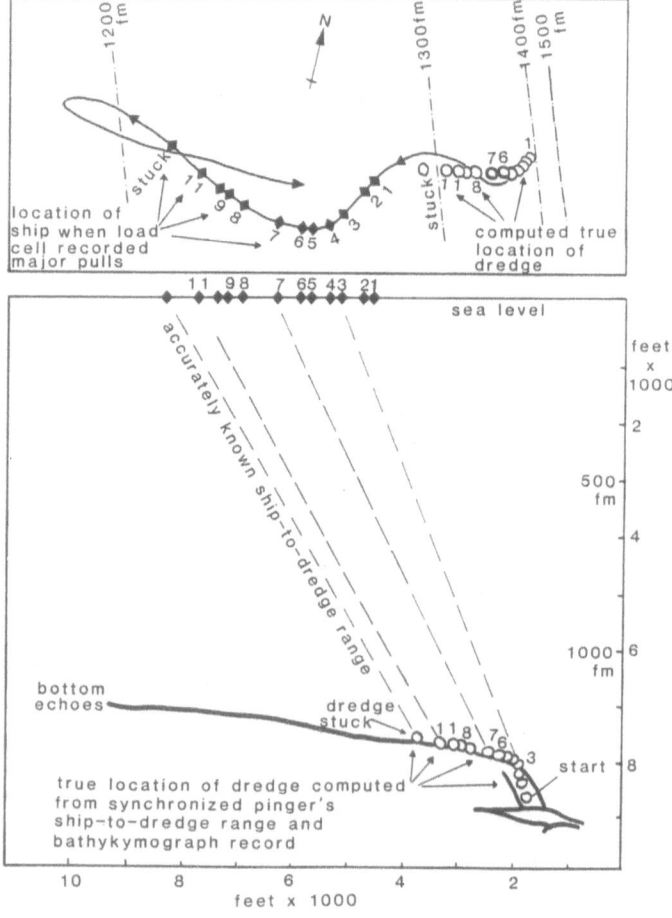

Fig. 7. (above) Plan view of a dredge run showing the estimated "bite" positions. (below) Elevation view showing how the "bite" times are converted into estimated positions (from Aumento, 1970).

ship's warp as it passed over the afterdeck in order to "feel" the "bites" as the dredge was supposedly sampling!

Through the early 1970s the most accurate deep-sea dredging operations relied on a system of moored radar transponder buoys, operated in conjunction with satellite navigation (Loncarevic, 1969) for control of the ship's position.

DREDGE "MONITORING"

Information on dredge behaviour in the early 1970s came from a ruggedly built "bathykymograph" (Aumento, 1970) or a pinger mounted 20 m ahead of the dredge to record both absolute depths reached by the dredge with time (actually ship-to-dredge range) and the instants and depths when it underwent the major "jerks" or abrasions. (These were also recorded by the load cell tensiometer on the ship.) Typical results from such an arrangement are compiled in Fig. 7.

Alternatively, a combination of a regular pinger, mounted up to 300 m from the end of the ship's cable to measure water depth, and a tilt-switch pinger (Fig. 4) mounted in a protective housing immediately ahead of the dredge was used for monitoring (Gaunt and Wilson, 1975). The latter pinger was designed so that its pulse-repetition rate changed

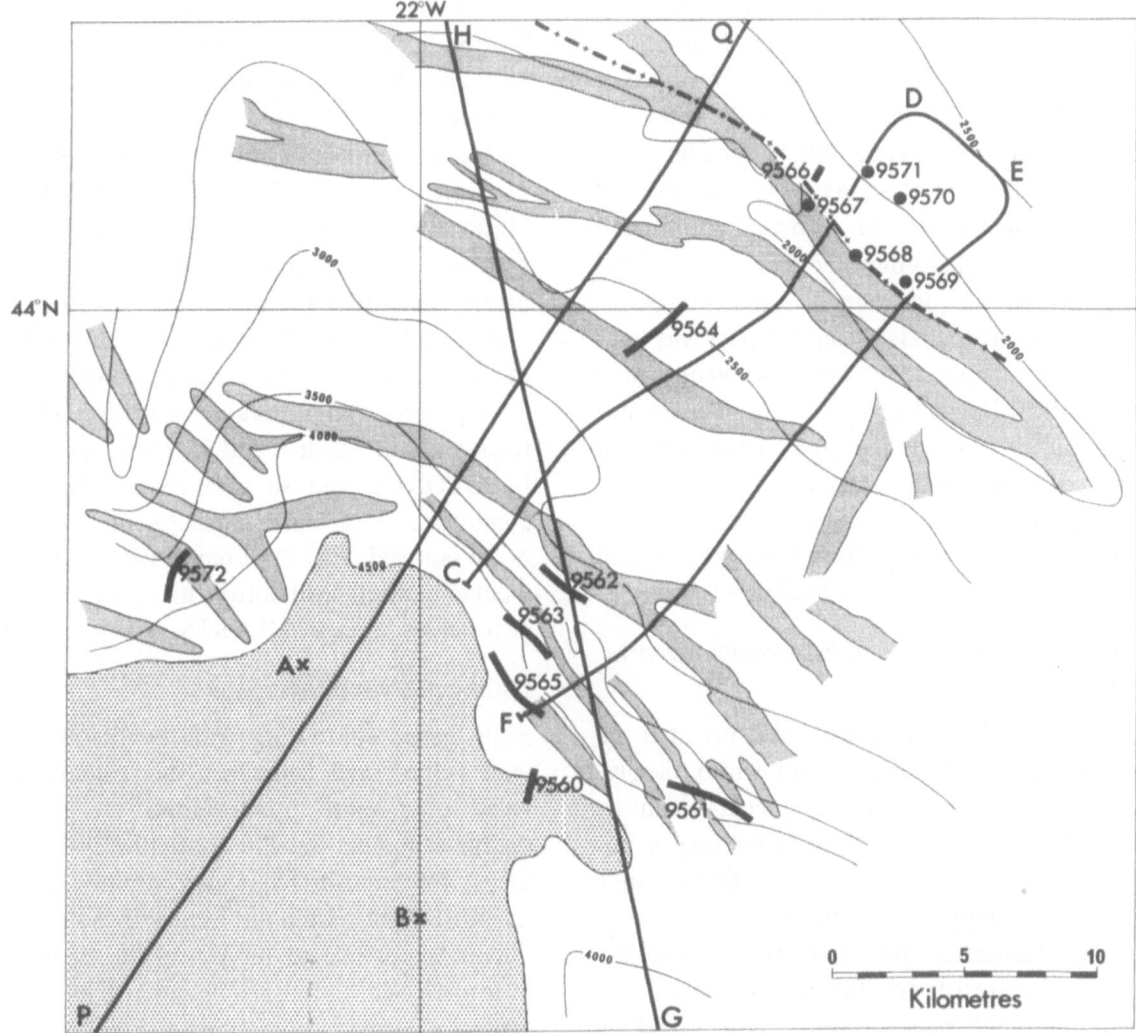

Fig. 8. North slope study area for the detailed sampling programme of "Discovery" Cruise 84. Contours are in corrected metres. Shaded areas represent fault scarp outcrops seen on GLORIA long-range sidescan monographs. Stippling shows the sedimented trough floor below 4500 m stations. Dots indicate core positions; bars represent dredge. Dashes and crosses show the trace of the summit of the ridge; A and B are the positions of the acoustic beacons. CDEF, GH and PQ are seismic profiles (after Kidd et al., 1982).

when its orientation (and that of the dredge) changed from vertical to horizontal. The acoustic signal from this pinger was received on the ship's hydrophone, and vertical and horizontal modes could be determined by changing the facsimile recorders' scan speed (Fig. 6). The main advantage of the tilt switch pinger arrangement was that it was no longer necessary to make the assumption from the regular pinger record that, provided the height of the pinger off the bottom is substantially less than its distance from the dredge, then the dredge must be in contact with the bottom. In deeper water it was possible to misread the facimile recorder. The practice was to observe the crossing of the direct signal and the bottom echo and to count the number of complete phases, stopping when the pinger suggested the dredge was on the bottom. With the pinger away from the dredge this exercise could be flawed. The tilt switch pinger was mounted immediately ahead of the dredge and it also indicated exactly when the dredge was horizontal and so able to sample.

Some researchers at this time also attempted to monitor their operations with a small bottom camera/flash unit mounted along the ship's cable. This would give important information on where in the dredge traverse the material being sampled was talus or potentially from outcrops (Aumento and Lawrence, 1968; Laughton *et al*, 1971). Most major developments in the design and deployment of dredging equipment had taken place by the mid-1970s. Improvements in dredging operations after that came largely from improved navigation and from major advances in survey equipment.

DREDGING WITH BOTTOM-TRANSPONDER NAVIGATION AND GLORIA

In the late 1970s bottom transponders began to be used to provide a baseline around which deeply-towed vehicles and sampling equipment could be navigated by triangulation of acoustic ranges between the ship and the bottom transponders. Later an interrogator pinger was incorporated on the ship's cable (Robertson, this volume). At about the same time, the British Institute of Oceanographic Sciences long-range sidescan sonar system (GLORIA) (Somers *et al.*, 1978) provided a facility which displayed the regional morphology and the texture of the sea floor and provided new opportunities for precise dredging activities. Satellite-navigated GLO-

RIA imagery, together with seismic profiles were used to detect outcrops. On sonar images over rugged terrain, fault scarps give rise to sharp linear echoes and shadows, outcrops of volcanic basement create more diffuse or chaotic echoes and sediment cover shows up as fields of uniform low-level backscattering of acoustic energy. Thus with GLORIA an accurate sketch map of the regional geology could be drawn and constrained by crossing seismic profiles. These satellite-navigated survey data can then be used to select target dredge sites and the positioning for bottom transponders to navigate the dredging operations.

Kidd *et al.* (1982) used two transponders laid about 10 km apart, and 4 km away from a scarp that was to be sampled as part of dredging operations in 1977 within the King's Trough tectonic complex in the Northeast Atlantic Ocean (Fig. 8). The transponders were on tethered lines about 200 m above the sea floor and were "surveyed in" with additional echo sounder and seismic profiles. An interrogator pinger was towed in a protective "fish" 200 m along the trawl warp ahead of the dredge. The make-up of the rest of the system followed that of the IOS rectangular dredge arrangement with tilt switch pinger as described by Gaunt and Wilson, 1975 (Fig. 4). This first attempt at transponder-navigated dredge sampling in King's Trough provided accurate navigation of the ship relative to the baseline and to the scarps as identified by GLORIA, and so provided a high degree of confidence as to which outcrops and terraces were being sampled (Fig. 8). It paved the way for development of more efficient bottom navigation of the dredge gear across areas of GLORIA survey in later years.

DREDGING WITH SEABEAM:

Bathymetric contouring with swathe-mapping systems (Grant and Schreiber, this volume) has provided even more opportunities for precision in dredge sampling. Satellite-navigated swathe contour maps, computer-drawn at a resolution of 10 m or less for rugged terrain are frequently so reproducible that it is possible to reposition the operating ship over a selected fault scarp or outcrop and accurately control a dredging operation to sample it. This simplified mode of operation has become increasingly popular during the 1980s with French and American institutions who operate hull-mounted Seabeam systems. It avoids the time taken to deploy bottom

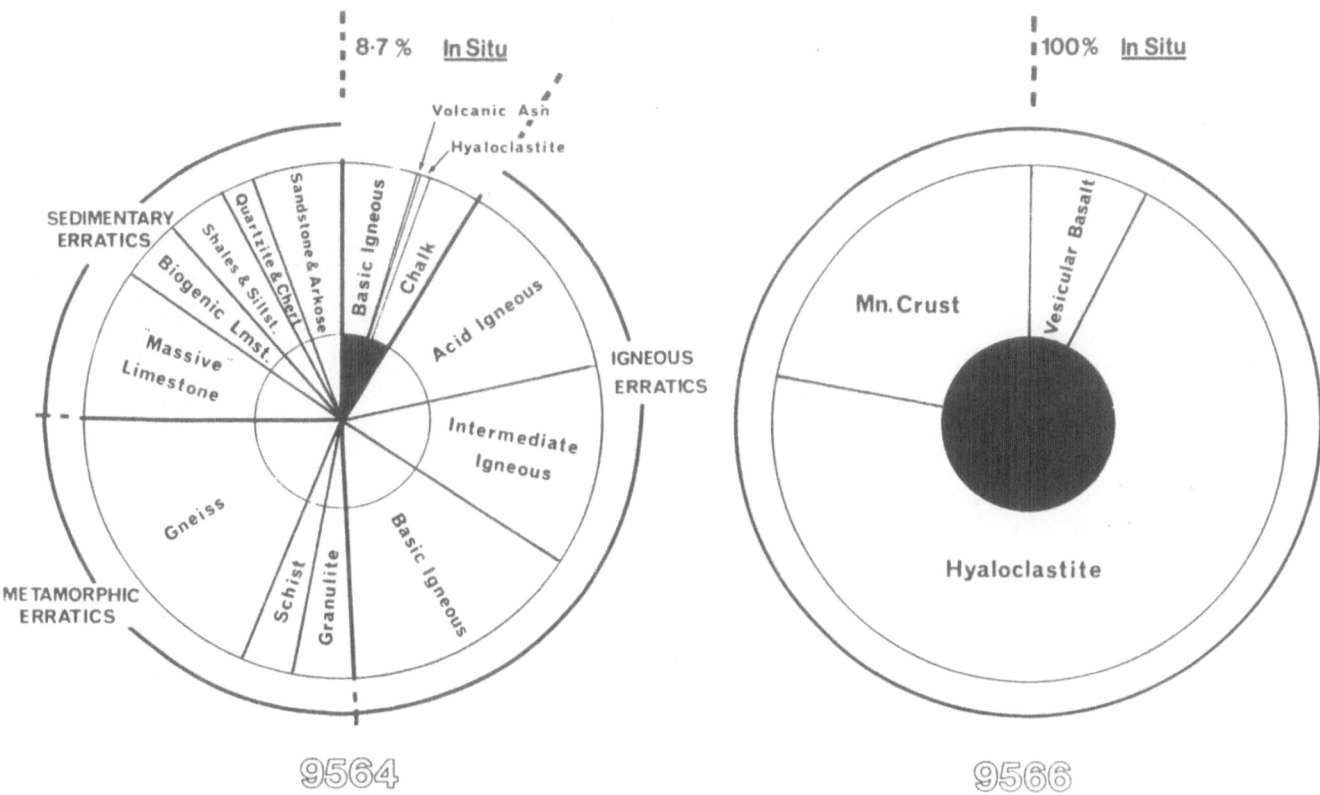

Fig. 9. Classification of material from two dredge hauls from the North Atlantic showing the variability of ice rafted content of dredge material (after Huggett and Kidd, 1984).

transponders and to survey in their baseline, which can run to a full day of shiptime.

More advanced versions of long-range sidescan and swathe bathymetry systems are coming into use as the decade closes. An advanced GLORIA system with quantitative bathymetry is mooted (M. Somers, 1989, pers. comm.) utilizing principles of signal measurement similar to those developed in Seamarc I and its derivatives. Hydrosweep (Grant and Schreiber, this volume) and other derivatives of the Seabeam systems are now operational. All are likely to have a profound effect on any future advances in dredging technique, since all these survey tools are destined to be used in conjunction with the Global Positioning System (GPS) (Robertson, this volume). Already the order-of-magnitude improvement in ship's navigation provided by GPS, to within 1 m accuracy, can mean that, if dredging is carried out over GLORIA or Seabeam-surveyed terrain, the operator has considerable control over which scarps are being sampled, even in the absence of a bottom transponder net.

Dredge Haul Recovery

There are a number of factors that may affect the amount and type of recovered material in an individual dredge haul.

IN SITU AND EXOTIC MATERIAL

The types of recovered material are broadly divided into the "*in situ*" and the "exotic". *In situ* rocks are either freshly-broken samples from outcrop or are rocks recovered from talus or scree accumulating at the bases of scarps. Exotic material on the ocean floor principally comes from floating ice (glacial erratics) and is much less confined to high latitudes than had been previously thought. Generally, hauls at higher latitudes than 40° should now be *expected* to contain glacial erratics (Kidd and Huggett, 1981), but investigators should note that erratics have been recovered from seamounts within 30° of the equator. Other exotic material can be transported to the sea floor attached to help hold fasts (Emery and Tschundy, 1941), by sea mammals (Emery, 1963), as

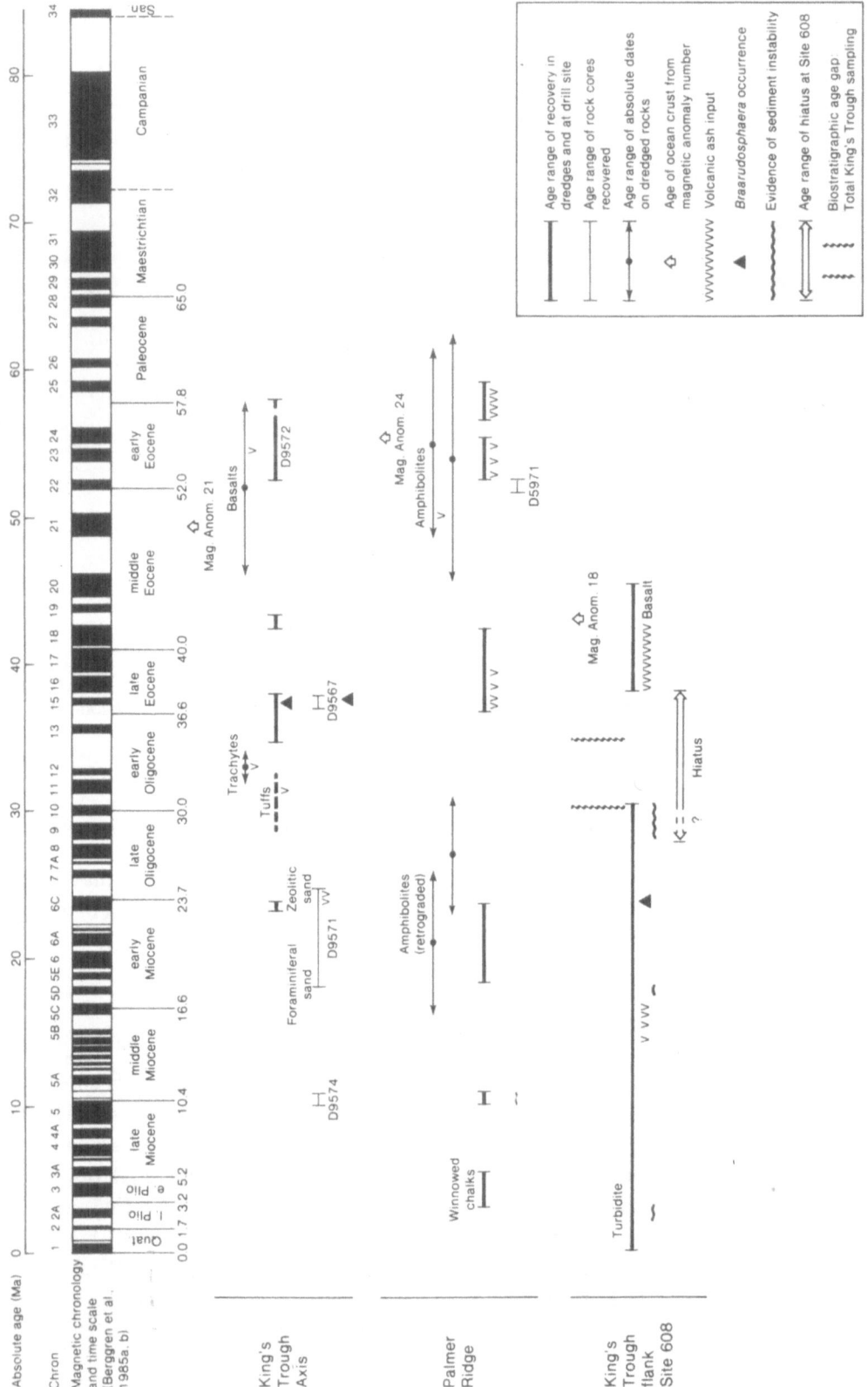

Fig. 10. Comparison of the age ranges of dredged materials from two North Atlantic study sections (Kings Trough and Palmer Ridge) with the drilled stratigraphy at a Deep Sea Drilling Project site (DSDP site 608) from the same area. Ages for rock cores taken in the dredging locations are also shown. Note that the dredging recovery partly encompasses a time gap represented by a Hiatus at Site 608. (See key for further detail).

ballast and clinker discharged from sailing and streamships, as modern man-made litter or as floating volcanic pumice. Thus location of the sampling area in relation to glaciated land areas, past and present shipping lanes, volcanically active areas and sea-mammal migration routes can be important in determining the content of a dredge haul.

Huggett and Kidd (1983/84) have constructed a set of criteria for the often-difficult identification of glacial erratics in dredge hauls. The criteria are in two categories. The first includes observations based on direct evidence visible on individual clasts, for example glacial striations and facetting. The second category includes inferred evidence drawn from complete dredge hauls, for example, variations in the thickness of manganese coating and groupings of the rocks into assemblages. Typical results of this approach on two North Atlantic dredge hauls are shown in Fig. 9.

DREDGE DESIGN AND OPERATIONS RECOVERY

As outlined in the Section, *Types of Dredges*, dredge design and operations probably have the most significant effects on haul recovery. The size and shape of the dredge mouth will control maximum clast size collected, while the size of links in the collecting bag or of any finer mesh netting liner will control minimum clast sizes retained. Although it is assumed that most dredges collect all the loose material they cross that is small enough to enter the mouth, some dredges are designed to ride over (e.g. the rectangular dredge of Nalwalk *et al.*, 1961) or sideways (e.g. the cylindrical dredge of Filippov *et al.*, 1970) past obstructions. Not all dredges have substantial cutting teeth that are capable of breaking outcrop (like the rectangular dredge of Aumento, 1970). With exceptionally large hauls some dredges may lift off and spill some of their contents, thus it is sometimes difficult to link haul size to distance travelled over the bottom (Kidd and Huggett, 1981).

SHIP'S TRACK RECOVERY

Careful survey and planning of ship's track during dredging can improve the quality of material recovered. Scree-like accummulations of talus at the foot of fault scarps can result in large hauls but these are also the sites that, in high latitudes, commonly produce high proportions of glacial erratics (up to 45%). Precise dredging around bottom transponders

has also highlighted the effect of ship's track in faulted terrain. Hauls taken along-strike frequently contain more material but this is dominated by manganese-coated talus and erratic material as compared to the small hauls taken across strike that can contain a high proportion of freshly broken material.

Comparison with Drilling

Prior to the advent of scientific deep ocean drilling in 1968, geological studies of sea floor features were heavily dependent on dredging. Typically, an area or feature would have been extensively surveyed geophysically with echo-sounding, seismic reflection profiling, gravity, magnetics and seismic refraction before detailed station work with rock coring, bottom photography and dredging. The deep-sea drilling vessel *Glomar Challenger* allowed investigators to precisely test interpretations of the geophysical data and to calibrate seismic sections by drilling reflectors at depth. With steady improvement in drilling technique through the Deep Sea Drilling Project and, since 1986, through the Ocean Drilling Program aboard DV *JOIDES Resolution*, a continuously-cored section that establishes the stratigraphy of an area became possible using HPC + XCB + RCB drilling (see Storms, this volume) and this has become the standard. Although dredging is frequently used during site survey to establish the scientific case for drilling a particular feature, the technique has become recognized as a poor substitute for a drilled section.

Dredging cannot provide the investigator of basement petrology with entirely unaltered material and this has provided a strong case for carrying out drilling on basement features that may be suspected as being composed of easily altered material such as serpentinite in diapirs (Boillot *et al.*, 1987; Kastens *et al.*, 1987).

A study which was able to directly compare the results of dredging and rock coring with drilling data and which favours dredging was reported by Kidd and Ramsay (1986). A drilling site occupied late in the DSDP programme at King's Trough in the north east Atlantic was designed to test a hypothesis on the formation of this tectonic complex that had been put forward based on a range of geological and geophysical data including GLORIA sidescan and bottom transponder navigated dredging (Kidd *et al.*, 1982).

A continuously cored section, DSDP Site 608, was drilled into Eocene oceanic basement at 515.4 m sub-bottom on one of the flanks of the tectonic complex, penetrating reflectors that were identified on regional seismic profiles. One of the reflectors is represented by an Eocene/Oligocene hiatus of around 9 m.a. in duration. All of the rock types recovered in the drilling had already been recognized and dated in dredge hauls. Some of the dredge materials spanned part of the period of hiatus development and some haul material from the east of the complex pre-dated the rocks immediately above the basement at the drill-site (Fig. 10). Kidd and Ramsay (*op. cit.*) were able to use the drill-site as a reference stratigraphic section and to combine this with the dredge haul data and thus refine the tectonic model and timing of formation of this intraplate feature.

The major features of the model were, however, originally defined on the basis of a series of cruises that conducted dredging and, although an early drill-site may have resulted in the same conclusions, some of the information on timing and the character of regional tectonic events would not have come available without a number of sites being drilled. We conclude that dredging can be a useful additional tool, even when the drilling of sequences is attainable. Additional regional information might be derived by extension of the geology away from a drilled reference section using dredging. There may be many locations like Kings' Trough that have been drilled but would benefit from at least a fresh analysis of regional dredges hauls.

Comparison with Submersible Studies

Fox and Stroup (1981) stress that there is no substitute in complex mid-ocean ridge terrain for close-up geological observation using submersibles. However, the amounts of selected samples that can be collected in submersible dives are very limited. Boillot *et al.* (1987) very successfully combined initial dredge haul results on the Galicia continental margin with both deep-sea drilling and follow-up submersible observations. Clearly the general availability and the expense involved in both submersible operations and drilling mean that many investigators may have to confine their efforts to dredge activity; however, the combination of all three types of opera-

tion is likely to provide the most powerful investigative approach.

The Future

Many marine geologists are recognizing, at the beginning of the 1990s, a move towards conducting deep-ocean geological studies at the outcrop scale rather than solely through regional remotely-sensed investigations. We have the "Landsat and aerial photography" level tools in the derivatives of GLORIA and Seabeam with which to select sea floor sites and set about studying the geology at the outcrop level. With the precision afforded by GPS and bottom transponder nets dredging probably will experience an upsurge of interest, at least in reconnaissance studies of newly explored regions. Subsequent detailed studies of such areas, however, are likely to be conducted by remotely-controlled combined camera and sampling systems (e.g. rock drills), submersibles and/or by dynamically-positioned and video-monitored drilling. On the other hand, we can also foresee dredging being used to extend information gained at reference drillsites.

References

Aumento, F., 1970, Improved Positioning of Dredges on the Seafloor, *Can. J. Earth Sci.* **7**, 534–539.

Aumento, F. and Lawrence, D. E., 1968, Photographic Control of Deep Sea Dredging, *Geol. Surv. Can. Paper* **68** (9), pp. 1–3.

Boillot, G., Winterer, E. L., and Meyer, A. W. *et al.*, 1987, *Proc. ODP 'Initial Reports' 103*: College Station TX (Ocean Drilling Program) 809–828.

Burt, R. G., 1979, Handling Oceanographic Equipment: Notes and Sketches from a Netman's Log, *Institute of Oceanographic Sciences Report* **34**, 62 pp.

Cronan, D. S., 1980, *Underwater Minerals*, Academic Press, London, 362 pp.

Emery, K. O., 1963. Organic Transportation of Marine Sediments, in M. N. Hill (ed.), *The Sea*, Vol. 3. Wiley and Sons, New York, pp. 776–789.

Emery, K. O. and Tschundy, R. H., 1941, Transportation of Rock by Kelp, *Geol. Soc. Amer. Bull.* **52**, 855–862.

Fillippov, L. A., Krausch, A., Barash, M. S., Laurov, V. M. and Dmitriyea L. V., 1970, A Large Cylindrical Dredge, *Oceanology*, 140–142.

Fox, P. J. and Stroup, 1981, The Plutonic Foundation of the Oceanic Crust, in Emiliani, C. (ed.), *The Oceanic Lithosphere*, Vol VII, The Sea: Ideas and Observations on Progress in the Study of the Sea, J Wiley and Sons, New York, pp. 119–218.

Gaunt, D. I. and Wilson, J. B., 1975, Acoustic Monitoring of Dredge Behaviour on the Sea Floor, *Deep Sea Res.* **22**, 91–97.

Huggett, Q. J. and Kidd, R. B., 1983/84, Identification of Ice-Rafted and Other Exotic Material in Deep Sea Dredge Hauls, *Geo-Marine Letters* **3**, 23–29.

Kastens, K. A., Mascle, J., Auroux, C. *et al.*, 1987, *Proc. ODP Initial Reports 107*: College Station TX (Ocean Drilling Program).

Kidd, R. B. and Huggett, Q. J., 1981, Rock Debris on Abyssal Plains in the Northwest Atlantic: A Comparison of Epibenthic Sledge Hauls and Photographic Surveys, *Oceanologica Acta* **4**, 99–104.

Kidd, R. B. and Ramsay, A. T. S., 1986, The Geology and Formation of the King's Trough Complex in the Light of Deep Sea Drilling Project. Site 608 Drilling, in Ruddiman, W. F., Kidd, R. B., Thomas, E. *et al.*, *Init. Repts. of Deep Sea Drilling Project*, Vol. 94, Washington, pp. 1245–1261.

Kidd R. B., Searle, R. C., Ramsay, A. T. S., Pritchard, H., and Mitchell, J., 1982, The Geology and Formation of King's Trough, Northeast Atlantic Ocean, *Marine Geology* **48**, 1–30.

Kinoshita, Y., Maruyama, S., Honza, E., Yamakado, N., Usami, T., and Handa, K., 1975, Technical Notes on Deep Sea Bottom Sampling, in Mizuno, A. *et al.*, *Cruise Report No. 4. Deep Sea Mineral Resources Investigation in the Eastern Central Pacific Basin. Geol. Surv. Japan*, 49–61.

Laughton, A. S. *et al.*, 1971, RRS Discovery Cruise 33, *N.I.O. Cruise Report No 33*, National Inst. of Oceanography, Wormley, Surrey, 23 pp.

Loncarevic, B. D., 1969, Buoy Plot as a Survey Aid. Trans. Applications of Sea-Going Computers Symposium, *Mar. Tech. Soc.*, Washington D.C., pp. 27–33.

Nalwalk, A. J., Hersey, J. B., Reitzel, J. S., and Edgerton, H., 1961, Improved Techniques of Deep-Sea Rock-Dredging, *Deep Sea Res.* **8**, 301–302.

Somers, M. L., Carson, R. M., Revie, J. A., Edge, R. H., Barrow, B. J., and Andrews, A. G., 1978, GLORIA II—An Improved Long-Range Sidescan Sonar, *Proc. IEEE/IERE Sub-Conference on Offshore Instrumentation and Communications*, OCEANOL. INT. 1978, Tech. Sess. J. BPS Publications Ltd., London, pp. 16–23.

The Use of Sediment Traps in High-Energy Environments

J. WHITE

Department of Civil Engineering, University of Southampton, Southampton S09 5NH, UK

(Received 27 April, 1989; 1 September, 1989)

Key words: sediment traps, high-energy environments, suspended sediment sampling.

Abstract. A sediment trap is a container deployed in the water column with the aim of providing a representative sample of the material settling through that water column before it passes to a greater depth and ultimately to the seabed or lake bottom. A review of the previous literature shows cylinders and baffled funnels to be the most efficient sediment trap design in flows less than 0.1 m/s. For flow velocities above 0.1 m/s recent evidence suggests upwelling from the trap base, and possible undercollection. The degree of undercollection depends on the flow velocity, the type of trap, the height: diameter (aspect) ratio of the trap, and the type of sediment. Recent experiments suggest that cylinders with an aspect ratio of $\geqslant 3$ may be efficient collectors in velocities up to 0.2 m/s. The use of cylinders is not recommended in velocities above 0.2 m/s. For unbaffled asymmetric funnels a lower limit of 0.12 m/s is suggested.

Introduction

Over the past fifteen years sediment traps have become an increasingly popular tool for investigating particulate flux in oceanic and lacustrine environments. The aim of a sediment trap is to provide a representative sample of the material settling through the water column, before it passes to a greater depth and ultimately to the seabed, or lake bottom. Most of the early sediment trap studies were undertaken in environments in which the current velocities were below 0.1 m/s. Laboratory experiments have verified that in such conditions certain trap designs provided an accurate estimate of the vertical flux. The subsequent deployment of sediment traps in a wider range of conditions, such as the continental slope, submarine canyons, estuaries, and the nearshore zone, meant that assumptions about trap performance were made beyond the hydrodynamic conditions for which they had been tested.

Only recently have experiments been conducted to assess sediment trap behaviour in flow velocities above 0.1 m/s. This chapter provides an assessment of sediment trap designs, and reviews recent developments in the use of traps in high-energy environments.

Sediment Trap Shape

Since the first recorded use of a sediment trap by Heim (1900) a variety of designs have evolved to suit individual needs, the designs commonly being based on intuitive assumptions of trap behaviour rather than tested models. Gardner (1980a) classified the designs into the five broad categories below:

  i) Cylinders;
  ii) Funnels;
  iii) Wide-mouthed jars;
  iv) Containers with bodies much wider than the mouth (e.g. Flasks and Tauber Traps);
  v) Basin/tray-like containers with width much greater than height.

The different shapes are illustrated in Fig. 1.

The earliest studies to investigate the particulate flux values determined from traps of different shapes were simple field comparisons. Pennington (1974) found that the sedimentation rate inferred from cylindrical traps deployed in Lake Windermere agreed closely with known rates from core samples, palaeomagnetic evidence and Pb^{210} dating. Funnel traps deployed simultaneously, however, tended to give sediment accumulation rates of 0.3 to 0.5 of the expected value (i.e. to "undercollect" sediment). This supported the earlier work of Johnson and Brinkhurst (1971) who reported differences in the collection efficiencies of cylinders and funnels deployed in the

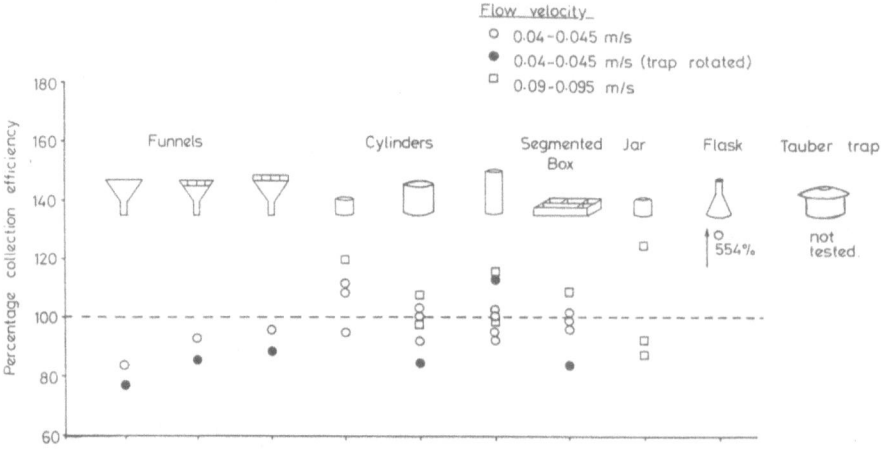

Fig. 1. The collection efficiency of different sediment trap shapes (adapted from Gardner, 1980a).

Bay of Quinte and Lake Ontario which appeared to depend upon the size of the funnel or cylinder used. Tauber traps tested by Pennington (*op. cit.*) tended to "over collect" sediment at an average of 2.3 times that collected in cylinders. When compared against basin/tray traps however, Reynolds and Godfrey (1983) found Tauber traps in Lake Windemere to collect up to 25 times more sediment. In the oceanic environment, Dymond *et al.* (1981) observed only a factor of two variation in the amount of sediment collected by cylindrical, funnel and basin/tray traps in the Santa Barbara Basin.

It is only through laboratory investigations that a reliable measure of the response characteristics of different types of sediment trap has been obtained. The turning point in sediment trap methodology came with the work of Gardner (1977), later summarized in Gardner (1980a and b). Models of the five different sediment trap shapes were tested in a recirculatory flume for flow velocities in the range 0–0.095 m/s. The trapping efficiency was calculated by comparing the sediment flux measured in the trap (mass/cm² of trap opening/per unit time) with the sedimentation rate on the flume bed. The results (Fig. 1) show cylinders, segmented boxes and baffled funnels to be the most efficient trap shapes. The effect of rotating the traps through 180°, 45°, and 135° during the experiments, to simulate a change of current direction is also shown in Fig. 1. Since the work of Gardner (1977) subsequent laboratory tests, e.g. those of Hargrave and Burns (1979) and Butman (1986), together with other reviews of existing data, such as those of Bloesch

and Burns (1980), Reynolds *et al.* (1980) and Blomqvist and Hakanson (1981) have recognized cylinders to be the most efficient sediment trap shape.

The experiments of Butman (op. cit) have verified that in flows up to 0.1 m/s baffled funnels also provide a good estimate of particulate flux. Baffled funnels have been used extensively in deep oceanic environments, for example by Honjo (1980) in the Sargasso Sea and E. Hawaii Abyssal Plain and Jickells (1984) also in the Sargasso Sea. Funnels have the distinct advantage of concentrating the collected material in a sample container at the funnel base from which resuspension is unlikely during retrieval.

Sediment Trap Size

The earliest investigations into the effect of sediment trap size on collection efficiency considered different sediment trap designs. Davis (1967) found in laboratory experiments that the amount of material collected in cylindrical jars with openings varying between 25 and 100 cm², was directly proportional to the area of the trap mouth (Fig. 2). Field experiments using funnels by Watanabe and Hayashi (1971) in a lake environment yielded similar results. These results indicate that for fixed relative dimensions of the trap, i.e. height: diameter ratio or "aspect ratio", the cross-sectional area of the trap will not affect the amount of sediment per unit area.

The collection efficiency of traps is significantly affected, however, if the relative dimensions (aspect ratio) are changed. Most of the work in this field has

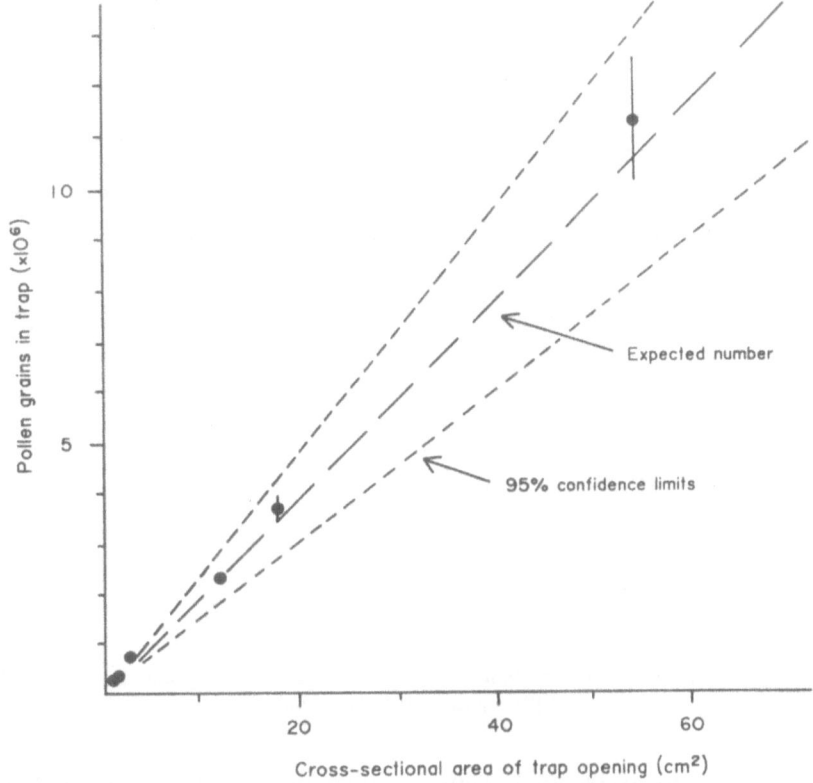

Fig. 2. Relationship between amount of material collected and trap cross-sectional area (adapted from Davis, 1967).

been concerned with defining the optimum aspect ratio for cylinders in any given hydrodynamic region. The flume experiments of Gardner (1980a) over the velocity range 0–0.095 cm/s showed the aspect ratio to have no apparent influence on the trapping efficiency for cylinders with aspect ratios of 1, 1.1 and 2.3. In field conditions near the Woods Hole Oceanographic Institution, however, where the velocity occasionally reached 0.5 m/s, an increase in the material collected was observed with increasing aspect ratio, and Gardner (1980b) suggested an optimum aspect ratio of between 2 and 3. This relationship was also investigated in the laboratory by Hargrave and Burns (1979) at velocities of 0.04–0.05 m/s for aspect ratios of 1.2, 2.6, 3.6, 5 and 20.4 and by Blomqvist and Kofoed (1981) in the Baltic Sea for ratios of 0.5, 1, 2, 3, 4, 6, 8 and 10. The results indicate that the apparent flux rate of material into the trap (i.e. the amount of material trapped per unit area, per unit time) increases with the aspect ratio up to a value of about 3 (Blomqvist and Kofoed, 1981) or 5 (Hargrave and Burns, 1979). After this the flux rate tends to a constant value

which depends upon the prevailing hydrodynamic regime (Fig. 3).

Cylindrical sediment traps are thought to collect sediment by:

(i) Particles falling directly into the trap; and

(ii) Particles being carried into the trap by trap-induced turbulence.

The asymptotic relationship between collection efficiency and aspect ratio is thought to mark the dominance of the process of particles being carried into the trap by trap-induced turbulence. The critical aspect ratio marks the point at which a quiescent zone is formed at the base of the trap. Above this aspect ratio there is very little change in the amount of material collected; below this limit however, eddies may resuspend material from the trap base.

The Relationship between Trap Collection Efficiency and Flow Velocity

All the aforementioned work was conducted in flows less than 0.1 m/s. The encouraging results from sediment trap deployments in these environments lead to

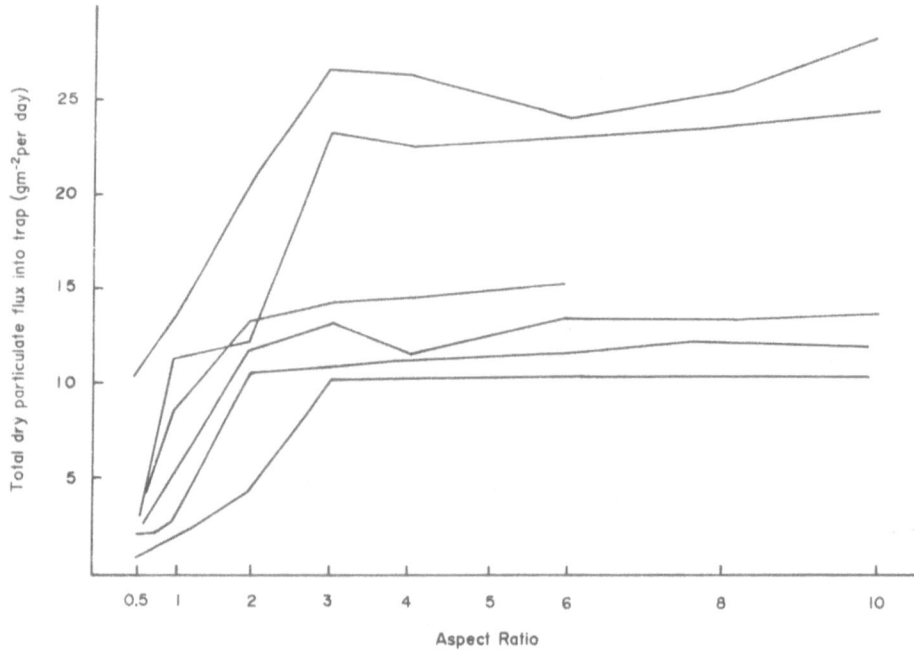

Fig. 3. Evidence for asymptotic relationship between amount of material collected in trap and aspect ratio, for cylindrical traps (after Blomqvist and Kofoed, 1981).

their use in a wider range of hydrodynamic conditions. Parmenter *et al.* (1983) used traps in flows with a mean speed of 0.3 m/s off Georges Bank, Gardner *et al.* (1983) deployed traps in the "Hebble" area where mean current speeds ranged between 0.08 m/s and 0.32 m/s, and recently Gardner (1989) investigated resuspension in the Baltimore Canyon where the mean velocities were up to 0.19 m/s, with maximum velocities of 0.8 m/s. It is only very recently however, that the relationship between collection efficiency and flow velocity has been fully examined.

The first investigation of the relationship between trap efficiency and aspect ratio under a wider range of hydraulic conditions was undertaken by Lau (1979), who considered the aspect ratio, h/d, in relation to the trap Reynolds number R_t, Ud/v, where;

h = height of trap
d = diameter of trap mouth
U = velocity of fluid at trap mouth
v = kinematic viscosity.

In a series of flume experiments the motion of oil droplets at the trap base was observed over velocities between 0.03 and 0.75 m/s in cylinders with aspect ratios of between 4.7 and 10 i.e., a range of R_t values between 2×10^3 and 3×10^4. By observing whether

the oil droplets stayed or escaped from the traps, Lau (*op. cit.*) determined the aspect ratio at which upwelling would occur for any given hydrodynamic conditions (Fig. 4). Unfortunately, the range of conditions and aspect ratios tested by Lau (*op. cit.*), and the use of oil droplets rather than sediment particles, limits the extent to which these results can be applied to natural sedimentary environments.

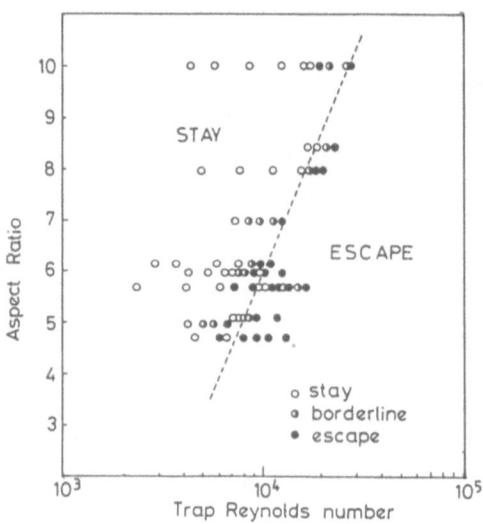

Fig. 4. The influence of aspect ratio and trap Reynolds number on oil droplet movement from a cylinder's base. Line indicates stay/escape boundary (after Lau, 1979).

Subsequent experiments examining the relationship between hydrodynamic conditions, aspect ratio and collection efficiency have used the R_t value to characterize the flow. In flow visualization experiments Gardner (1985) observed a tranquil zone at the base of cylindrical sediment traps with aspect ratios of 5, in velocities of up to 0.22 m/s. The corresponding R_t value was 8.4×10^3, which compares favourably with the value of 8×10^3 in Lau's experiments. Recent flow visualization experiments by Hawley (1988) have shown that upwelling of a layer of dye at the trap base in cylinders with aspect ratios of 5 starts at $R_t = 4.9 \times 10^3$, and is almost continuous at $R_t = 8.5 \times 10^3$.

At lower aspect ratios the upwelling in cylinders occurs at lower R_t values. The results of Butman (op. cit.) suggest that cylinders with aspect ratios of 3, accurately collect sediment at $R_t = 2.2 \times 10^3$, but at 4.6×10^3 significantly less sediment is collected (Fig. 5). Hawley (op. cit.) has shown that in a cylinder with an aspect ratio of 3, upwelling starts at $R_t = 3.5 \times 10^3$ and is almost continuous at $R_t = 5.1 \times 10^3$.

Information regarding the performance of other trap designs at velocities above 0.1 m/s is scant. For R_t values of 1.0 to 1.2×10^3 Butman (op. cit.) showed that wide-mouth jars overcollected sediment and funnels undercollected sediment as compared to cylinders, supporting the earlier work of Gardner (1980a). In the same experiments, baffled funnels and cylinders collected similar amounts.

Recently Baker et al. (1988) tested an unbaffled asymmetric funnel over a range of velocity condi-

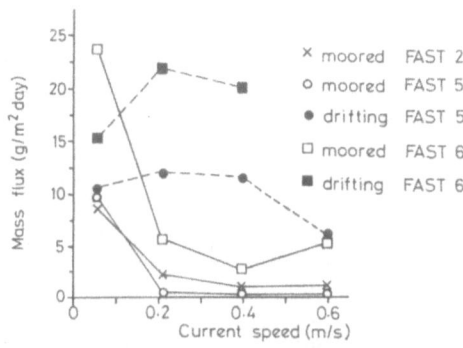

Fig. 6. Amount of material collected in asymmetric funnels at different velocities (from Baker et al. 1988).

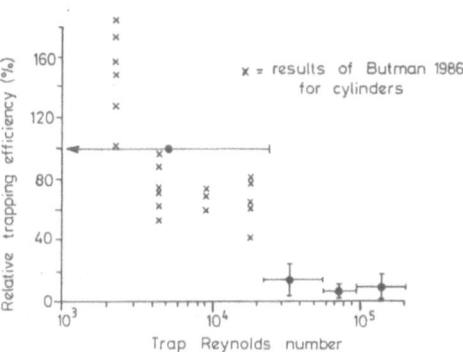

Fig. 7. Decrease in relative trapping efficiency with increasing trap Reynolds number for an asymmetric funnel (from Baker et al. 1988).

tions in Colvos Passage, Puget Sound. A Flow Activated Sediment Trap (FAST) was developed, capable of partitioning the collections according to the velocity regimes in which the collection occurred. The velocities were <0.12 m/s, 0.12 – 0.3 m/s, 0.3 – 0.5 m/s, and 0.5 m/s. The results were compared to similar free drifting sediment traps deployed simultaneously, considered to give an accurate estimate of the vertical particle flux since there is little velocity shear across the trap mouth. Figure 6 shows the results of the moored traps and the free drifting traps plotted against the velocities. Clearly less sediment is collected at velocities above 0.12 m/s in the moored traps. Taking the collections at 0.12 m/s to represent 100% efficiency Baker et al. (op. cit.) have shown the drastically reduced efficiency of asymmetric funnels at higher R_t values (Fig. 7).

Laboratory Experiments of Trap Efficiency versus Velocity

As a precursor to the deployment of cylinders in an estuarine environment where the current velocities

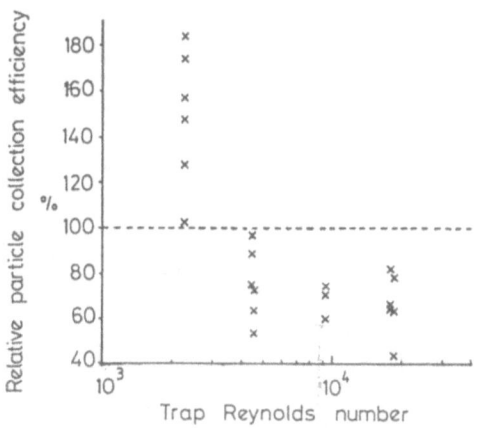

Fig. 5. Relative particle collection efficiency vs trap Reynolds number for cylinders with aspect ratios of ~3 (from Butman, 1986).

TABLE 1

Dimensions of the cylinders tested in laboratory calibrations

External diameter (ED) (mm)	Internal diameter (ID) (mm)	Internal height (mm)	Aspect ratio Internal (height/ diameter
100	94	282	3
100	94	188	2
100	94	194	1
75	69	276	4
75	69	207	3
75	69	138	2
75	69	69	1
50	44	220	5
50	44	176	4
50	44	132	3
50	44	88	2
50	44	44	1

reached up to 0.4 m/s White (1989), conducted a series of laboratory experiments to investigate the collection efficiency of 12 cylinder types (Table I) at velocities of 0.1, 0.2, 0.3, and 0.38 m/s. The laboratory experiments were conducted in a 22.5 m long, 1.37 m wide and 0.6 m deep recirculatory flume. The freshwater of the flume was "seeded" with natural sediment taken from Port Hamble Marina, Hamble UK, to a concentration of approximately 60 mg/1. For each experiment, 24 cylinders were tested simultaneously, as shown in Fig. 8 with the mouth of each cylinder at 0.3 m above the bed in a water depth of 0.52 m. Before the start of each run the sediment was stirred into suspension within the flume by producing

a current of 0.38 m/s in the channel and sweeping the entire length of the channel bed with a weighted domestic broom. The mean mid-channel velocity was adjusted to the desired setting, monitored with an electromagnetic current meter, and the cylinders were placed within the flume.

Concentration profiles were measured at either end of the test section, at the beginning and end of each run. Since there were no closed spaces in the system, and it was assumed that the flow velocity in the pumps and pipes was too great for particles to settle, the difference between the two sets of concentration values gives a measure of the total amount of sediment to have settled either on the flume bed or in the traps.

Each trap was removed after 24 hours, the trapped sediment was filtered through a preweighed glass microfibre filter, dried at 105°C, desiccated, and then weighed to determine the dry sediment weight in the trap. This was then corrected for the amount of material remaining in suspension within the trap at the time of its retrieval, to specify the total dry weight of sediment collected on each trap base. This was then compared to the calculated amount of material collected on the flume bed per unit area. For a 100% efficient trap in any given flow, the amount of sediment collected on the trap base and that collected on the flume bed (per unit area) would be equal. The procedure was similar to that performed by Gardner (1980a).

Figures 9a and b show the collection efficiencies plotted against velocity for cylinders with different aspect ratios at 0.1 and 0.2 m/s respectively. The

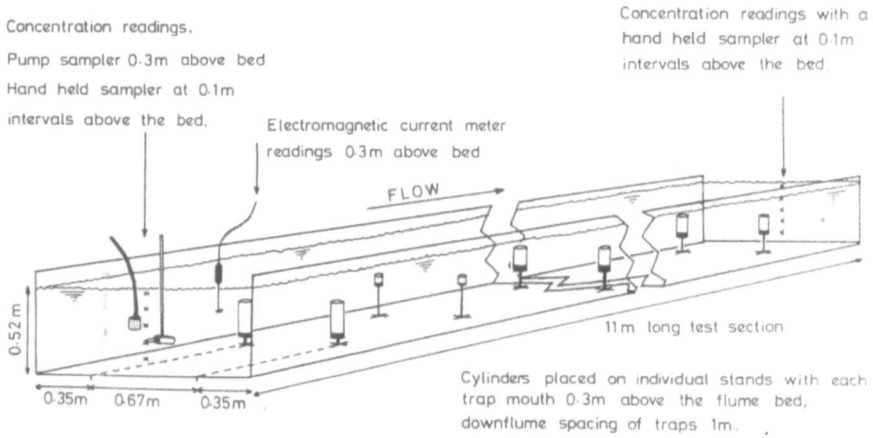

Fig. 8. Flume layout for experiments testing the effect of velocity on the collection efficiency of cylinders.

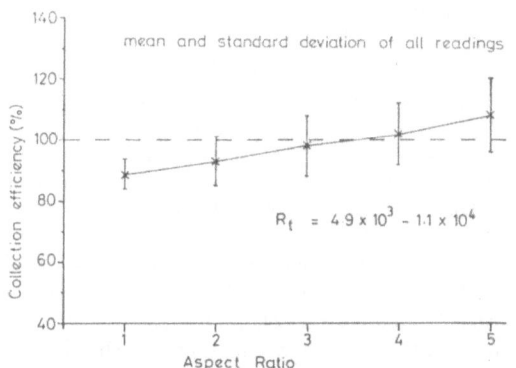

Fig. 9(a). The collection efficiency of cylinders with different aspect ratios at 0.1 m/s (from White, 1990).

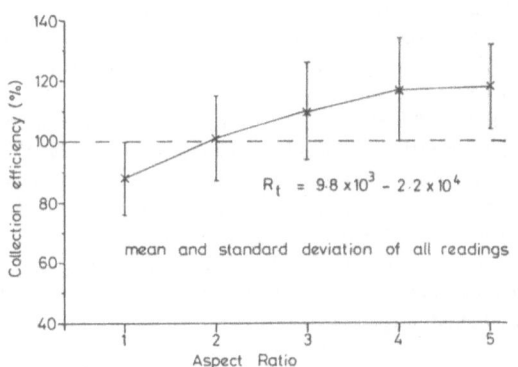

Fig. 9(b). The collection efficiency of cylinders with different aspect ratios at 0.2 m/s (from White, 1990).

results suggest that at velocities up to 0.2 m/s cylinders provide a reasonably accurate estimate of vertical flux.

The flow visualization experiments from previous literature suggest that for the range of R_t values tested (given on Figs. 9a, b) upwelling should have occurred from virtually all traps, particularly at 0.2 m/s. The fact that undercollection did not appear significant in cylinders with aspect ratios greater than 1 suggests that upwelling does not necessarily produce particle resuspension.

Butman (*op. cit.*) also found that traps with aspect ratios of 2.7 collected efficiently at $1 \times 10^4 R_t$, whereas Hawley's experiments suggest upwelling should be complete at $5.1 \times 10^3 R_t$ for this aspect ratio. Clearly further work is needed to verify the limit at which resuspension occurs rather than upwelling of fluid.

Unfortunately the results of White (1990) for flows of 0.3 and 0.38 m/s are inconclusive due to

resuspension of material from the flume bed at these velocities.

Conclusions

Laboratory experiments and field investigations have suggested that cylinders and baffled funnels are the most efficient sediment trap designs for estimating the vertical flux in velocities up to 0.1 m/s. At higher velocities (or R_t values) recent work suggests that upwelling from a trap base may occur although the point at which particle resuspension and hence undercollection occurs is still unclear. The point at which resuspension occurs depends upon the trap type, the trap aspect ratio, the ambient velocity, and the sediment type. Recent laboratory experiments suggest that cylinders with an aspect ratio of 2 may be efficient collectors in velocities up to 0.2 m/s. As a precaution it is suggested that an aspect ratio of at least 3 and preferably 5 is used in deployments in such environments. The use of cylinders in flows above 0.2 m/s is not recommended. Unbaffled asymmetric funnels have been shown to seriously undercollect sediment at velocities above 0.12 m/s.

Further work is needed to investigate resuspension of particles from the trap base, and any biasing effects that resuspension may have on the composition of the particles collected. If sediment traps are to be used in high-energy environments such as the continental slope, estuaries, and the nearshore zone, the limitations outlined in this paper must be considered when the results are interpreted.

Acknowledgement

Figures in this paper are reproduced from the "Marina 89" Conference proceedings, published by Computational Mechanics Publications, UK, with the kind permission of the publishers.

References

Baker, E. T., Milburn, H. B., and Tennant, D. A., 1988, Field Assessment of Sediment Trap Efficiency under Varying Flow Conditions, *J. Mar Res.* **46**, 573–592.

Bloesch, J. and Burns, N. M., 1980, A Critical Review of Sedimentation Trap Technique, *Schweizerisch Zeitschrift für Hydrologie* **42**, 15–55.

Blomqvist, S. and Hakanson, L., 1981, A Review of Sediment Traps in Aquatic Environments, *Arch Hydrobiol*, **91**, 101–132.

Blomqvist, S. and Kofoed, C., 1981, Sediment Trapping—A

Subaquatic *in situ* Experiment, *Limnology and Oceanography* **26**, 585–590.

Butman, C. A., 1986, Sediment Trap Biases in Turbulent Flows: Results from a Laboratory Flume Study, *J Mar Res.* **44**, 645–693.

Davis, M. B., 1967, Pollen Decomposition in Lakes as Measured by Sediment Tarps, *Geol Soc Am Bull.* **78**, 849–858.

Dymond, J., Fischer, K., Clauson, M., Cobler, R., Gardner, W., Richardson, M. J., Berger, W., Soutar, A., and Dunbar, R., 1981, A Sediment Trap Intercomparison Study in the Santa Barbara Basin, *Earth and Planetary Sci Lett.* **53**, 409–418.

Gardner, W. D., 1977, *Fluxes, Dynamics and Chemistry of Particulates in the Ocean*, PhD Thesis MIT/WHOI Joint Program in Oceanography, pp. 405.

Gardner, W. D., 1980a, Sediment Trap Dynamics and Calibration: A Laboratory Evaluation, *J. Mar. Res.* **38**, 17–39.

Gardner, W. D., 1980b, Field Assessment of Sediment Traps, *J. Mar. Res.* **38**, 41–52.

Gardner, W. D., 1985, The Effect of Tilt on Sediment Trap Efficiency, *Deep Sea Res.* **32**, 349–361.

Gardner, W. D., 1989, Baltimore Canyon as a Modern Conduit of Sediment to the Deep Sea. *Deep Sea Res.* **36**, 323–358.

Gardner, W. D., Richardson, M. J., Hinga, K. R., and Biscaye, P. E., 1983, Resuspension Measured with Sediment Traps in a High Energy Environment, *Earth and Planetary Sci Lett.* **26**, 262–278.

Hargrave, B. T. and Burns, N. M., 1979, Assessment of Sediment Trap Collection Efficiency, *Limnology and Oceanography* **24**, 1124–1136.

Hawley, N., 1988, Flow in Cylindrical Sediment Traps, *J. Great Lakes Res.* **14**, 76–88.

Heim, A. 1900, Der Schlammabsatz am Grund des Vierwaldstatter see, *Vierteljahresschrift Naturforschenden Gesellschaft in Zurich.* **A5**, 164–182.

Honjo, S., 1980, Material Fluxes and Modes of Sedimentation in the Mesopelagic and Bathypelagic Zones, *J. Mar. Res.* **38**, 53–97.

Jickells, T. D., Deuser, W. G., and Knap, A. H., 1984, The Sedimentation Rates of Trace Elements in the Sargasso Sea Measured by Sediment Trap, *Deep Sea Res.* **31**, 1169–1178.

Johnson, M. G. and Brinkhurst, R. O., 1971, Benthic Community Metabolism in Quinte Bay and Lake Ontario, *J. Fisheries Resource Board of Canada* **28**, 1715–1725.

Lau, Y. L., 1979, Laboratory Study of Cylindrical Sedimentation Traps, *J. Fisheries Resource Board of Canada* **36**, 1128–1291.

Parmenter, C. M., Bothner, M. H., and Butman, B., 1983, Characteristics of Resuspended Sediment from Georges Bank Collected with a Sediment Trap, *Estuarine and Coastal Shelf Sci.* **17**, 521–533.

Pennington, W., 1974, Seston and Sediment Formation in Five Lake District Lakes, *J. Ecology* **62**, 215–251.

Reynolds, C. S., Wiseman, S. W., and Gardner W. D., 1980, An Annotated Bibliography of Aquatic Sediment Traps and Trapping Methods, *Freshwater Biological Assoc. Occasional Publication* **11**.

Reynolds, C. S. and Godfrey, B. M., 1983, Failure of a Sediment Trapping Device, *Limonology and Oceanography* **28**, 172–176.

Watanabe, Y. and Hayashi, H., 1971, Investigation on the Method for Measuring the Amount of Freshly Precipitating Matter in Lakes, *Japanese J. Limnology* **32**, 40–45.

White J., 1990, *The Use of Sediment Traps to Monitor Marina Siltation*, PhD Thesis, Civil Engineering Department, Univ. of Southampton (pending).

Pore Pressures in Marine Sediments: An Overview of Measurement Techniques and Some Geological and Engineering Applications

P. J. SCHULTHEISS*

Schultheiss Geotek, Fern Cottage, Marley Lane, Haslemere, Surrey GU27 3RF, UK

(Received 27 April, 1989; accepted 1 September, 1989)

Key words: in-situ measurements, pore pressure, marine sediments, measurement techniques, hydrology, sediment properties, effective stress, geotechnical.

Abstract. Pore pressures in the seabed are extremely sensitive to any imposed stress because of the low permeabilities commonly exhibited by marine sediments. Consequently, the measurement of sediment pore pressures can be used to infer either the nature of the imposed stress (if the sediment properties are known) or the physical properties of the sediment (if the imposed stresses are known). Stresses of many different types may be exerted on the seabed either through hydrostatic forces (e.g. tidal and wave effects), or directly by lithospheric forces (e.g. tectonic and thermal forces). Several techniques for measuring *in situ* pore pressures in the upper few metres of sediments have been developed, and one instrument, the PUPPI, will operate autonomously in water depths up to 6000 m. Basic sediment properties and processes can already be inferred from pore pressure responses using this technique. However, further application and development could greatly enhance its capability, especially for long-term monitoring of sediment conditions. In this Chapter, pore pressure measurement techniques are briefly reviewed and problems are highlighted. An outline is given of some of the many ways in which pore pressure measurements could be used to gain further insight into geological processes and to determine some of the pertinent sediment properties more accurately for engineering applications.

Introduction

The importance of knowing the pore pressures within a mass of soil, in order to understand the soil's behaviour, has been recognized for many years. Baligh (1986) succinctly summarizes this importance:

"Porewater pressures occupy a central position in modern soil mechanics for conceptual and practical reasons. Conceptually, effective stresses control most soil behaviour aspects of interest to geotechnical engi-

neers and total stresses are controlled by equilibrium conditions. Hence pore pressures are necessary to estimate effective stresses from calculated total stresses and thus to allow the rational interpretation and/or prediction of the response of soil masses. Practically, the pore pressure in the soil is often easier to measure than other equally meaningful aspects of soil behaviour because it exhibits no directional dependence. This is especially so in field situations involving anisotropic and non-uniform stressing (or straining) of the soil."

Not only is the pore pressure a fundamental soil parameter that has to be known if a soil or sediment behaviour is to be understood, it is also often extremely sensitive to stresses imposed on the soil. Laboratory testing of soils is often accompanied by detailed pore pressure measurements in order to understand the soil's behaviour, however, *in situ* pore pressure measurements are far less common and are even rarer in marine sediments. The high pore pressure sensitivity to imposed stresses in marine sediments arises because they are commonly saturated and have relatively low permeabilities. The measurement of sediment pore pressures can be used to infer either the nature of the imposed stress, if the sediment properties are known, or the physical properties of the sediment, if the imposed stresses are known.

In marine sediments the measurement of *in situ* pore pressures can be complicated by the relatively inaccessible environment and high ambient hydrostatic pressures. However, once these difficulties are overcome, the marine environment offers some distinct advantages for pore pressure measurements over the land environment. These include saturation of the sediments in most marine environments (this improves the response time of the measurement system),

* Previously at: Institute of Oceanographic Sciences, Deacon Laboratory, Wormley, Godalming, Surrey GU8 5UB, UK

important for very accurate measurements) and naturally occurring dynamic stresses (e.g. hydrostatic tidal cycles) that can be used in *in situ* experiments to determine sediment properties. Above all there is a host of applications for accurate pore pressure measurements in marine sediments in all water depths that could be used for strategic, commercial or purely scientific ends.

Pore Pressures

STATIC AMBIENT FORE PRESSURES

In order to understand the critical importance of pore pressure measurements to engineering and hydrogeological applications it is essential to appreciate the concept of effective stress as used in soil mechanics. Figure 1 is used to illustrate the changes in pressure and vertical stress within a water-sediment column. For simplicity, it is generally assumed here that the sediment beneath the overlying water column is fully saturated (no free gas) and that there are no density or permeability gradients within the sediment. It is also assumed that the total vertical stress, σ_v, at any depth is caused simply by the weight of the overlying sediment and water column.

The hydrostatic water pressure, U_h, increases linearly with depth in the water column (assuming constant density), and under conditions of static equilibrium (no movement of pore water) the water

pressure within the sediments (the pore pressure) is also linear and equivalent to hydrostatic pressure. However, the pore water pressure within the sediments, U, is not necessarily equal to hydrostatic (in Fig. 1, U is shown as being greater than the hydrostatic pressure). The differential pressure, ΔU (often called the "excess pore pressure" when it is greater than the hydrostatic pressure) is given by $U - U_h$.

The stresses that largely control the strength and deformation behaviour of unconsolidated sediments are the effective stresses. Terzaghi (1943) showed that the vertical effective stress, σ_v', is given by

$$\sigma_v' = \sigma_v - U.$$

ΔU and σ_v' are depicted in Fig. 1 which illustrates that the important effective stress is independent of the water depth.

If pore water is flowing vertically within the sediment column, then the pore pressures are not hydrostatic and are driven by the differential pore pressure, ΔU. This differential pressure gradient can be either positive (from an upward flow of pore water) or negative (from a downward flow of pore water). Assuming that the water flows according to Darcy's Law, then the seepage velocity, v (m s^{-1}) of an element of water in the soil is given by;

$$v = k \, \Delta U / \rho g n z,$$

where k is the hydraulic conductivity of the sediment (m s^{-1}), ΔU is the differential pressure (Pa) at a depth z (m), n is the fractional porosity, g is the acceleration due to gravity (m s^{-2}) and ρ is the density of the fluid (kg m^{-3}).

The consequence of this water flow and differential pore pressure gradient is to either increase the effective stress (from negative differential pore pressures) or to decrease the effective stress (from positive differential, or excess, pore pressures). It is therefore obvious that to determine the *in situ* state of the vertical effective stress, both the total stress and the total pore pressure need to be determined. The total vertical stress can be calculated from the density profile determined from a core sample (assuming that the total stress is simply related to the overburden) but the only direct method of determining the pore pressure is by *in situ* measurement.

There is a natural upper limit to the magnitude of excess pore pressure that can exist in an unlithified sediment and hence an upper limit to the seepage

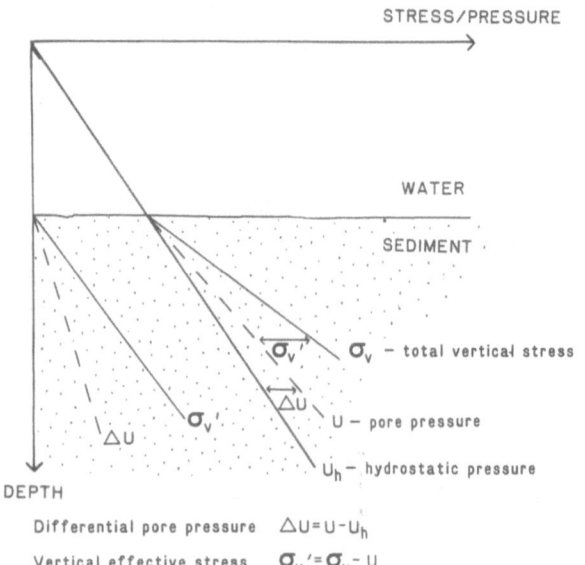

Fig. 1. Generalized profiles of pressure, total stress and effective stress in a water-sediment column.

velocity of pore water. Above this limit the sediment will liquefy and behave as a fluid. Cohesionless sediments (silts and sands) will liquefy when the effective stress is zero, i.e. when the pore pressure equals the total stress. In cohesive sediments (muds and clays) it is possible for small negative effective stresses to exist without the sediment liquefying. It should be noted that there is no corresponding maximum for downward advection velocity as negative pore pressure increase the effective stress and the sediment stability.

SEDIMENT PORE PRESSURES

Dynamic pore pressure effects in sediments can occur as a result of any stress wave. In this chapter only surface water waves are considered. They are of considerable interest and fall into two categories: standing waves and travelling waves. Only standing waves and very long-period travelling waves such as tides have any effect in the deep ocean because the dynamic hydrostatic pressure does not attenuate with depth. In shallow water, travelling surface waves can have a significant effect on the sediment pore pressure.

STANDING WAVES

Hydrostatic pressures from surface standing waves do not attenuate with depth; consequently their effects occur even in deep ocean sediments. Long

period surface gravity waves, such as tides, can be considered as standing waves, even in full oceanic depths (Hurley, 1989). The pressure amplitude response caused by surface standing waves in a water–sediment column is shown as a function of depth in Fig. 2a. This pressure response in the sediments is governed by the elastic properties of the sediment. If the sediment is fully saturated then the response is governed primarily by the sediment shear modulus, G, and the sediment permeability, k, as the other sediment parameters are essentially constant (Hurley, 1989). In Fig. 2a the limiting cases are for a sediment with zero rigidity ($G = 0$, i.e. a suspension) where the pore pressure is a maximum and remains equal to the hydrostatic pressure (U_h) and for an impermeable sediment ($k = 0$) where the pore pressure is a minimum U_{min}. A real sediment with finite values of permeability and shear modulus will have a pore pressure profile, U, which lies between these two limiting cases. Figure 2a is somewhat simplified as it assumes constant values of G and k with depth. The differential pore pressure, ΔU, measured between some point in the sediment and the overlying water is given by $U_h - U_{min}$.

TRAVELLING WAVES

The hydrostatic pressure caused by travelling surface waves attenuates exponentially with depth in a free

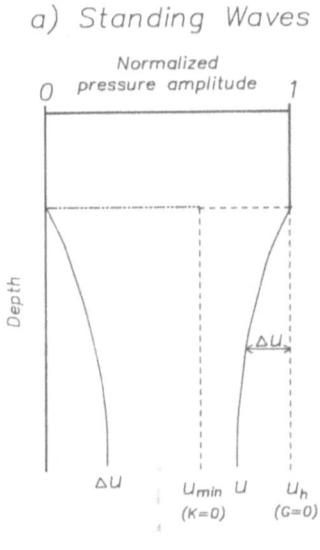

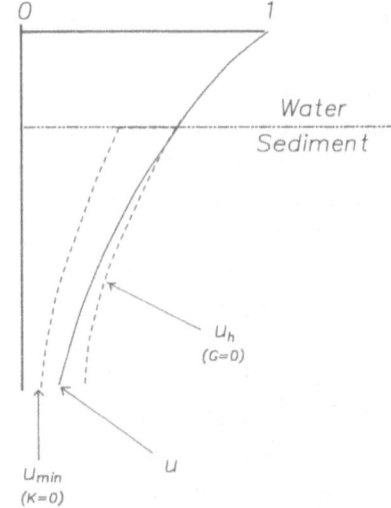

Fig. 2. Normalized pore pressure amplitude, U, in a water-sediment column caused by: a) standing waves and b) travelling waves. U_h and U_{min} are the limiting cases when the shear modulus, G, and the permeability, k, are zero.

water column. Consequently ocean waves, with periods of about 2–10 s, only influence sediments in relatively shallow water. Figure 2b depicts the pressure amplitude profile through the water–sediment column for surface travelling waves. The pore pressure in this case attenuates from two processes: (a) the exponential hydrostatic pressure decay, and (b) the decay caused by the elastic properties of the sediments and its permeability. It should be noted that a differential pore pressure measurement made between some depth in the sediment and at the sediment/water interface is not in this case ΔU ($U_h - U$). Therefore, the pressure attenuation caused by the sediment alone cannot be measured directly with a single differential pressure transducer.

Geological Applications

Hydrogeological processes in the ocean crust are major factors influencing the exchange of heat and elements between the crustal rocks and the oceans. Water circulating through the crust will influence not only the processes of plate tectonics but will also have a major effect on global geochemical budgets. A recent paper (Working Group 3, 1987) produced at the Second Conference on Scientific Ocean Drilling (COSOD II) clearly summarizes the importance of hydrodynamic fluid circulation within the crust for the further understanding of the mechanisms of crustal tectonics and the interrelationships, through the global geochemical budget, between the lithosphere, the hydrosphere, the atmosphere and the bio-sphere. Four primary areas of fluid circulation in the ocean crust were identified:

1. At the ridge axis spreading centre, where forced convection occurs at the ridge axis caused by large temperature gradients in the surrounding crust. It is this vigorous forced convection that causes the well known 350°C black smokers. At sediment-free ridge axes the low permeability sediment layer which inhibits water circulation at sedimented ridge axes is absent. As a result the residence time available for fluids to chemically react with the sediments and rocks is greater at sediment-covered ridge axes.

2. Free convective circulation is thought to occur on the ridge flanks, possibly between a few kilometres and as much as 1000 km away from the ridge axis. Sedimentary cover generally increases away from the ridge axis, increasing the resistance to con-

vective water flow. At the same time the heat flow from the crust decreases as it cools and more of the heat loss occurs by conduction rather than by convection. Heat flow surveys (e.g. Becker and Von Herzen, 1983) have indicated that this free convection may occur in cells that are in some cases evenly spaced and in other cases controlled by large-scale features such as faults. When the crust has cooled sufficiently and the resistance to fluid flow has increased, through both the decrease in permeability of the basement and the increasing thickness of sediments, the heat loss occurs only by conduction and major fluid circulation ceases. This is probably the general case for most ocean basins.

3. Hydrologic flow occurs at continental margins where continental rainfall raises the groundwater level above that of the surrounding seawater. Consequently, it can be expelled through aquifers directly into the ocean. This pressure-driven circulation can be enhanced by density-driven circulation where the groundwater passes through evaporites which raise the density to levels above that in seawater. Passive margins can also accumulate sediments of sufficient thickness to heat organic constituents and produce oil and gas. Fluid flow will occur upwards as a result of positive bouyancy of these products and by compaction of the thick layers of sediment.

4. Tectonically induced physical compaction of sediments occurs at active ocean margins, causing water to be expelled in the process of subduction and accretion. Although the patterns of pore water expulsion within accretionary complexes are not known in any detail, it is thought that dewatering at the toe of such complexes is pronounced (Westbrook and Smith, 1983). It is currently assumed that water is expelled through a variety of features such as thrust faults and mud volcanoes as well as in a pervasive manner through the bulk of the sediments.

These four hydrodynamic flow regimes are shown schematically in Fig. 3 (after Working Group 3, 1987), which also shows a very rough estimate of the possible total water discharge for each zone. It is interesting to note that while the ridge axes have the most vigorous circulation, the biggest contribution to the total discharge is from the ridge flank zone. Here the free convection is much slower but is cumulatively greater because of the much larger area over which it occurs.

In areas where pore water advection is taking

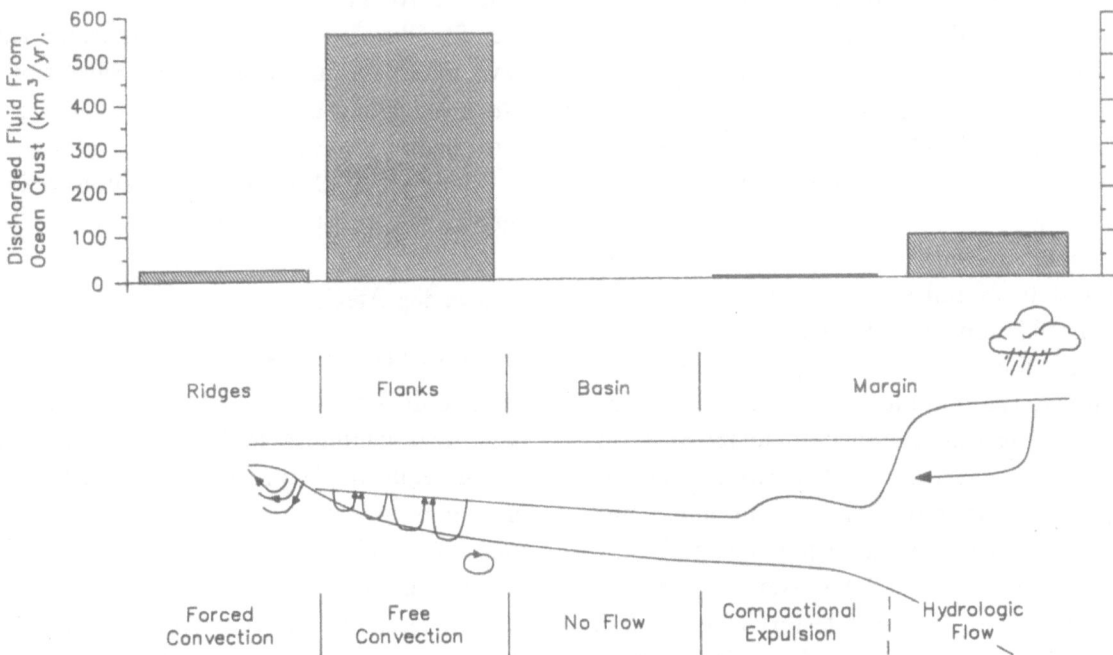

Fig. 3. Hydrodynamic zones and fluid flow processes, with estimates of total global discharge rates, on and around an oceanic plate (after Working Group 3, COSOD II).

place by a slow pervasive flow through the whole of the sediment body, *in situ* pore pressure and permeability measurements in the surficial sediments will enable the flow patterns to be defined and flow rates quantified. In some areas it is likely that the flow is much more vigorous, causing significant amounts of water to be either entering or undergoing expulsion from the sediments and ocean crust through faults, mud volcanoes or other small-scale features within the sediments. Accurate pore pressure measurements in the upper 10 m of sediments in these types of areas could still enable patterns of water movement to be identified within these hydrogeologically active areas. However, they would only be able to quantify the pervasive component of the overall flow pattern and other measurements or assumptions would still be needed to estimate the total flux of water through a given area.

Ocean Basins and Abyssal Plains

The ocean basins and abyssal plains are defined hydrogeologically by Working Group 3 (1987) as regions of diffusive flux where pore fluid convection in the oceanic crust has ceased. However, the transition between ridge flank and ocean basin is a gradual

transformation. It is not inconceivable that pore pressures other than hydrostatic do exist in ocean basin sediments and that the driving force may not be thermally induced but may be caused by other processes.

It was noted by Richards (1984) that pore pressures in excess of hydrostatic have always been measured *in situ* in cohesive soils, having a grain-size smaller than most sands at depths below a metre or two into the seabed. This generalization was based on only a small number of *in situ* piezometer measurements, primarily in soils on the continental shelf. Subsequent pore pressure measurements in the Madeira Abyssal Plain (Schultheiss and McPhail, 1986; Schultheiss and Noel, 1987) have shown that pore pressures in this region are generally hydrostatic. Nevertheless, there is some evidence to indicate that excess pore pressures may exist (at least to some depth) in many marine sediments as a result of the sedimentation and compaction processes themselves.

It is now generally accepted that many cohesive surficial sediments in the upper 4 m exhibit an apparent overconsolidation at high porosities. Apparent overconsolidation is not caused by the removal or erosion of any overlying material and has been attributed to the effects of high interparticle bonding, ageing and cementation, which result in an intrinsic strength sufficient to prevent normal consolidation and resembling an overconsolidated state in the laboratory (e.g. Hamilton, 1964; Richards and Hamilton, 1967; Noorany and Gizienski, 1970, Silva and Jordan, 1984).

The technique normally used to evaluate the consolidation state of soil samples is the one dimensional consolidation test on high-quality samples. From the $e - \log \sigma'_v$ (void ratio-log vertical effective stress) curves, the maximum past preconsolidation stress (σ'_c) can be estimated and compared with the calculated *in situ* vertical effective stress assuming hydrostatic pore pressures (obtained by integrating the sediment density to the required depth). The overconsolidation ratio (OCR) is given by σ'_c / σ'_v and is an indication of the stress history/consolidation state of the sediment (Silva and Mairs, 1987). OCRs around unity indicate a normally consolidated sediment, whereas values significantly greater or less than unity indicate over or under consolidation respectively.

Richards (1984) discusses the overconsolidation ratio profile found in many marine sediments which often decreases from a value greater than unity in the upper few metres to unity, or significantly below unity, as the depth increases. This underconsolidation at depth in sediments which have apparently been deposited continuously or episodically without any erosion over long periods of time, appears to be real and can be explained by the existence of excess pore pressure within the sediments. A recent example of a detailed study investigating the stress history of sediments in the Nares Abyssal Plain is that of Silva and Mairs (1987), who showed that the OCRs were less than unity between 5 m and 25 m below the seabed with minimum values of around 0.4 at about 15 m below the seabed. It would seem from this study that if the hypothesis of excess pore pressures is used to explain the data, then using a simple one-dimensional model of fluid flow requires that the excess pore pressures must exist not only in the deeper sediments, where the OCRs are less than unity, but also in the upper few metres, where the sediments may exhibit an apparently overconsolidated state. It is clear that to test this hypothesis *in situ* pore pressure measurements are needed in the same area. It is interesting to note that in the

Madeira Abyssal Plain, where pore pressures have been found to be generally hydrostatic, the sediments have OCRs of just over unity (Silva and Mairs, 1987), indicating a normally consolidated or apparently slightly overconsolidated state. However, there is no evidence of underconsolidation which is in agreement with the pore pressure data.

Engineering Applications

TOXIC WASTE DISPOSAL

Oceanic disposal of all kinds of waste products has been, and continues to be, considered and used. Ocean sediments have been considered as the primary barrier to the release of high-level radioactive waste (HLRW) into the ocean waters as part of an internationally co-ordinated programme investigating the feasibility of sub-seabed disposal (see Francis, 1984 for references). For safe burial of wastes into deep-sea sediments it is necessary to obtain a detailed understanding of the physical and geotechnical properties of the host sediments for the emplacement phase of the waste. This can only be fully achieved with *in situ* measurements, among which *in situ* pore pressure measurements are an important part as they can provide strength, consolidation and permeability data. In addition to the short-term site assessment requirements, the long isolation periods required for HLRW (tens to hundreds of thousands of years) requires that a detailed knowledge of any naturally occurring pore water flow patterns is obtained. The most direct method of determining pore water flux rates is by measuring any naturally occurring pore pressure gradients and the sediment permeabilities. It has been estimated that flow rates as low as 1 mm/year need to be resolved for safe disposal of HLRW. Pore pressure measurements in fine-grained sediments are capable of an accuracy in that order in fine grained sediments where the permeability is very low (Schultheiss and McPhail, 1986).

SITE INVESTIGATION AND SLOPE STABILITY

As seabed exploration and exploitation moves into deeper waters, foundation problems associated with slopes and different sediment types may be encountered. Slope failure may become a major hazard. Some site investigation techniques may change, with more emphasis placed on *in situ* measurements. Pore pressure measurements will undoubtedly figure prominently because of their fundamental importance to soil behaviour.

Submarine slope instability probably occurs over the entire spectrum of offshore environments from shallow, near shore zones to the continental slopes and beyond to the deep ocean floors (Prior and Coleman, 1984). Slope failures have been documented in offshore petroleum areas worldwide (notably in the Mississippi Delta), and must be reckoned with in any seabed construction or extraction operation, particularly in areas with moderate-to-high sedimentation rates, and where significant seabed slopes or gassy sediments prevail.

While it is known that sediment failures occur on even very gentle submarine slopes (less than 1°) the precise cause of the instability, or the cause and timing of the failure, is rarely known. In addition, there are currently no proven methods for evaluating the susceptibility of a specific area to likely failure or for monitoring changes in seabed conditions that may lead to failure.

One of the primary problems associated with an analysis of the slope stability of sedimented areas of the sea floor is an evaluation of the *in situ* effective stress. To estimate this, it is essential to measure *in situ* pore pressures (a component of the effective stress) within the sediments to the depth of interest, as well as the density profile (obtained from sampling).

It can be assumed that some types of sediments may be relatively unstable as a result of excess pore pressures. In fact all slope failures are probably initiated as a result of an increase in pore pressure. Ambient excess pore pressures can be caused by, for example, fast sedimentation rates, gas production or by naturally occurring advection caused by thermal gradients. For a failure to occur, the pore-pressure has to exceed a critical level. This critical level may be exceeded through either these slow changes in the ambient conditions or rapid transient causes such as severe storm conditions or seismic activity. The identification of areas at risk is of considerable

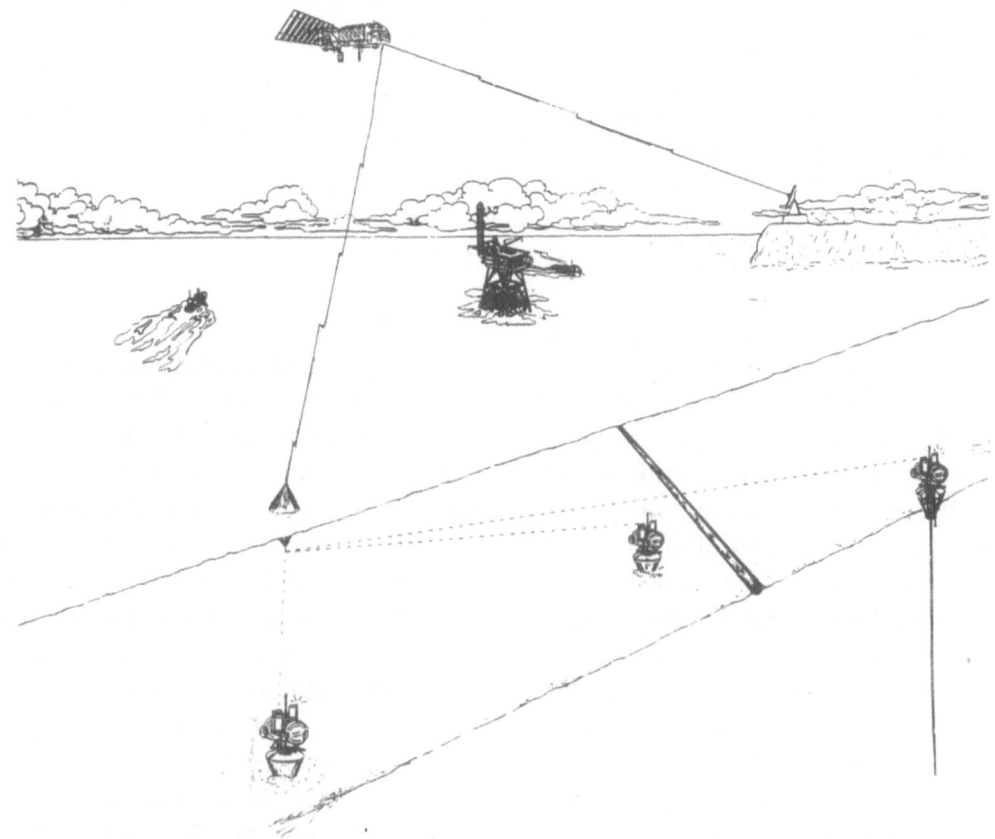

Fig. 4. Impression of a real-time monitoring system of pore pressures in the seabed. Data are transmitted acoustically from the sea floor to a data buoy and then via satellite to a monitoring station.

importance to any seabed engineering project. This is especially true as seabed exploration and hydrocarbon extraction extends into deeper waters and into areas that have significant slopes.

Future investigations into slope stability may require pore pressures to be monitored over long periods of time, to look for example at gas build-up in the sediments, changing pore pressure from rapid deposition of sediments, seasonal effects and/or the effects of storms. Indeed in some areas it is conceivable that real-time monitoring of pore pressures may be required purely for safety purposes. A schematic diagram of how this can be achieved, even in very deep or very remote areas is shown in Fig. 4. Pore pressure instruments could be strategically located in and around the area of interest or potential danger. The data would be transmitted to shore, rig or ship anywhere in the world via an underwater acoustic data-link to a surface bouy and then to a satellite system. The technology for this type of system already exists, and it would not be difficult to implement such a system if the need was perceived.

Review of Pore Pressure Measurement Technology

The potential value of having *in situ* pore pressure data from submarine sediments for engineering purposes has been recognized for many years. More recently, an awareness has developed of the importance of the role of fluid flow in many geological processes. Despite these facts very few instruments have been built to make these measurements and there is generally a lack of reliable data. Some of the instruments that have been successfully built and used will be briefly discussed here. This review is restricted to those devices that remain static in the seabed. It excludes, therefore, a number of devices that measure pore pressures as part of another test such as the piezocone test. It also excludes some piezometers that have been buried at very shallow depths to measure the effects of surface waves only.

The first known reported uses of a submarine piezometer were by Lai *et al.* (1968) and Richards *et al.* (1975). These authors describe a differential piezometer, built at the Norwegian Geotechnical Institute (NGI) and the University of Illinois (UI) that was designed to operate in water depths of up to 500 m. This NGI–UI probe was deployed several times in 1967 in the Wilkinson Basin, Gulf of Maine,

with one successful test being made at a water depth of 278 m. The probe had an overall length of 4.9 m and weighed 570 kg in air. It used a differential pressure transducer constructed using a boudon tube and vibrating wire with an accuracy of $+/-6.2$ kPa. A maximum excess pore pressure of 59 kPa was measured 3.2 m below the sea floor, caused by the insertion of the probe, which decayed to 9.8 kPa after 5 hours and did not decay significantly further after a total of 10 hours.

Submarine piezometer technology and data aquisition were most significantly advanced as a result of the SEASWAB experiments. SEASWAB (Shallow Experiment to Assess Storm Waves Affecting the Bottom) was one element of the US Geological Survey's broader study of sea floor instability known as the "Delta Project" and both of the SEASWAB experiments took place in an area known as "East Bay" in the Mississippi Delta. SEASWAB I (Garrison, 1977) was conducted in a water depth of 19 m, whereas SEASWAB II (Hottman *et al.*, 1978) took place 1830 m to the northeast of the SEASWAB I location in a water depth of only 13 m.

The piezometers used during the SEASWAB experiments were specifically designed for shallow-water applications. They took advantage not only of the shallow water but also of the nearby oil production platforms to directly record the pressure data via undersea cables connected to the instruments on the bottom.

SEASWAB I

The two piezometers used during the SEASWAB I experiments were on the sea floor for six months in 1975/76, and consisted of a large instrument built and operated by the National Oceanic and Atmospheric Administration (NOAA) and a smaller instrument, similar to the NGI-UI piezometer, operated by Lehigh University. NOAA's piezometer was 17.12 m long and constructed from 120 mm diameter pipe. It used variable reluctance, absolute pressure transducers to measure both the pore pressure at two locations (8 m and 15 m below the mudline) and the hydrostatic pressure at two locations (1 m and 15 m below the mudline). The Lehigh piezometer (Hirst and Richards, 1977) was about 7 m long and constructed from 54 mm diameter pipe. It used interconnected, fluid filled, bourdon tubes connected to vibrating wire strain gauges as a differential pressure transducer to measure the pore pressure at a depth of 6.4 m below the mudline. The piezometers were installed about 6 m apart by attaching weights and then lowering to the bottom by a crane. After installation the weight sections were removed with the aid of divers.

The details of the pore pressure data from the SEASWAB I experiment are not easy to explain fully. Hurricane Eloise passed over the site two days after insertion. This was probably prior to the complete dissipation of the insertion pressures and undoubtedly complicated the interpretation of the data. However, based on data extrapolation it was estimated that ambient excess pore pressures were 11.7 kPa and 44.1 kPa at depth of 8.4 m and 15.2 m respectively below the mudline from the NOAA piezometer (Bennett, 1977; Bennett *et al.*, 1976 and 1982). The data from the Lehigh piezometer indicated an ambient excess pore pressure of 32 kPa at 6.4 m below the mudline.

One of the primary problems encountered from the piezometer measurements made during SEASWAB I was identifying the effect of free gas on the pressure measurements in the sediments. Did the ambient pressure measurements determine the pore water pressure, the pore gas pressure or a combination of both? Did the contribution of water and gas pressure change with time? How much gas had migrated into the pore pressure measuring system? What was the effect of this gas on the dynamic pressures recorded? Did it simply increase the response times of the pressure measuring system to short-period pressure transients (wave effects) or were there other effects? All the porous pressure ports used during SEASWAB I were relatively coarse porous corundum stones (porosity = 40–50%, permeability = $1–3 * 10^{-2}$ cm/s) (Bennett and Faris, 1979).

SEASWAB II

SEASWAB II had basically the same objectives as SEASWAB I but an effort was made to increase the number of sensors and to measure separately both the pore water pressure and the pore gas pressure (Dunlap *et al.*, 1978). In an attempt to measure pore water pressure, fine-pore, high air entry (HAE) stones were used (porosity = 35–38%, permeability = $1–3 * 10^{-7}$ cm/s). It is generally accepted that the use of small pore-size filters as pressure ports

inhibits, by capillary effects, direct communication with free pore-gas (Bennett and Faris, 1979). Consequently, the pressures measured through the corundum stones may contain a contribution from high pore-gas pressure, whereas those pressures measured through HAE stones should not (Vaughn, 1973). The Lehigh piezometer was similar to that used previously except that it used a high air entry stone and a semiconductor-type pressure transducer. The NOAA piezometer was also similar to the one used previously except that it used a long slender tip to study pore water dissipation, and a different combination of transducers was used. Coarse stone corundum pressure ports were used at depths of 6.5 m, 12.6 m and 15.6 m below the mudline and an HAE pressure port was used at 15.6 m below the mudline. The transducers were all absolute sensors apart fom the one at 15.6 m with the corundum stone, which used a differential pressure transducer.

In addition to the Lehigh and NOAA piezometers, a third piezometer was installed during the SEASWAB II experiment. This piezometer (TAMU-USGS) was similar in construction to the NOAA instrument except that it used thin-film strain gauge transducers to measure pore gas pressure through corundum pressure ports at depths of about 3 m and 12.5 m below the mudline and to measure pore water pressure using an HAE port at about 12.5 m below the mudline.

Overall, the pressure data from the SEASWAB II experiment is confusing for several reasons. Both the Lehigh and the TAMU-USGS piezometers were located in a deformational feature and needed driving into the sediments which may have caused some damage to the instuments and hampered the determination of initial insertion pressures. The deformational feature is also believed to have moved significantly during a storm in late February 1977. Both of these piezometers sank into the mud during the course of the experiment; the TAMU-USGS instrument sank approximately 1.6 m and the Lehigh instrument sank approximately 1.8 m. The Lehigh piezometer malfunctioned for most of the time because of a cable problem, and on recovery it was found that the HAE stone had cracked and sediment was filling the cavity leading to the pressure transducer. The NOAA piezometer was installed outside the deformational feature and was free of these problems. On recovery it was found that there was hardly any fluid left in the cavities behind the porous stones irrespective of whether they were corundum or high air entry (Dunlap et al., 1978). The NOAA instrument provided the best data, showing high excess pressures increasing with depth for all the corundum ports (11 kPa at 6.5 m to 43 kPa at 15.6 m). However, the data obtained from the transducer connected to the HAE port at 15.6 m were erratic. A few days after insertion the excess pressure increased dramatically to a level to that recorded by the transducer with the corundum port at the same level. This was interpreted by Dunlap et al. (1978) as being caused by gas slowly diffusing through the HAE stone. However, after 40–50 days this pressure dissipated back to a much lower level and Bennett and Faris (1979) concluded that "further research is necessary to understand the significance of the large differences observed between HAE and corundum stones and the anomolous character of the HAE pore pressure record". The same data presented by Bennett et al. (1982) considered the HAE excess pressure data, which rose and fell during a period of about 65 days, to be anomolous. The NOAA instrument also provided data on the pressure response from surface activity. Pressures caused at depth within the sediment as a result of both tidal and wave activity were recorded (Bennett and Faris, 1979).

The NOAA piezometer was used again in 1978 in a water depth of 44 m in Block 73 of the Mississippi in an area known as "Main Pass" (Bennett et al., 1982, 1985). Pressure transducers were again located at depths of 6.5 m, 12.6 m and 15.6 m below the mudline with a combination of both corundum and HAE ports. Excess pore pressures measured at this site were considerably less than in the SEASWAB experiments in the East Bay area (14.5–17.9 kPa at 15.6 m and 15.8 kPa at 12.6 m). However, it is not clear whether there was less free gas in the sediments in this area or not. The insertion pressures were higher, which is believed to have been caused by the higher shear strengths of the sediments in the Main Pass area.

GISP

Another piezometer was built and used in shallow water in the Mississippi Delta by Sandia National Laboratories (Prindle and Lopez, 1983). GISP (Geotechnically Instrumented Seafloor Probe) is different from the piezometers described previously in that it is completely self-contained, stores the pressure data in an onboard solid-state memory and can telemeter these data acoustically to a surface vessel. GISP was designed to operate in water depths of up to 450 m. The Data Gathering System (DAGS) on the sea floor and the Command and Recording Subsystem on the surface (CARS) were developed for both the GISP and a Seafloor Earthquake Measuring System (SEMS) (Reece et al., 1979). The 10.5 m GISP probe was constructed from 108 mm diameter O.D. steel pipe containing absolute pressure transducers (quartz crystal, oscillating beam type) at 3.32 m, 6.42 m and 9.52 m below the mudline, with coarse corundum stones as pressure ports. A reference hydrostatic transducer was mounted 0.36 m above the mud plate. The GISP instrument, first tested in 1980, revealed several design problems. The modified GISP was tested in 1981 at the SEASWAB I site in a water depth of 19 m. Two identical probes were installed, 120 m apart, by lowering them into the sediment under their own weight (1725 kg) at a rate of 0.15 m/s in August 1981. The ballast weight (1180 kg) was lifted from the GISP instrument after installation by triggering a release motor. The two instruments were recovered in January 1982. Recovery was accomplished by acoustically releasing a bouyant reel of kevlar cable which was picked up and used to haul the whole instrument out of the mud and back to the surface. Peak insertion pressures of up to 63.4 kPa were recorded at 9.5 m. Despite some difficulties in interpreting the data it was concluded that ambient excess pore pressures increased with depth up to 17.9 kPa at a depth of 9.52 m below the mudline. This is in reasonable agreement with some of the data from NOAA piezometer used during the SEASWAB I experiments.

OXFORD UNIVERSITY

A differential piezometer has been built and used by Sills at Oxford University (OU) (Sills, pers. comm. and Sills and Nageswaran, 1984). The OU piezometer is constructed of stainless-steel pipe with a diameter of 44 mm, increasing at the top to 76 mm to house some instrumentationa. Ports are located at approxmately 2.3 m and 3.5 m below the mudline. The differential pressure transducers use a common, air-filled, back pressure reference system and have a high resolution (approx. 30 pa). Two absolute pressure transducers measure the common reference pressure and the seabed hydrostatic pressure. *In situ* pressure data have been obtained from three areas in shallow water around the UK, between 1978 and 1981, investigating primarily the effects of gas on the measured pressure response caused by tidal cycles. Tidal cycles with amplitudes of up to 14 m were monitored for periods of up to 60 hr. The data showed differential pore pressure amplitudes that ranged from 0% to about 12% of the tidal cycle at depths of up to 3.5 m below the mudline, indicating the presence of free gas in the sediments with an estimated degree of saturation between 97% and 98%. The OU piezometer is currently being used to investigate other sites around the UK.

BGI

A multiple piezometer was built by the Belgium Geotechnical Institute, BGI (Carpentier and Verdonk, 1986; De Wolf *et al.*, 1983) to measure pore water pressures caused by waves and tides in the foundation soils of a new outer harbour at Zeebrugge. The diameter of the piezometer probe varied from 60 mm at the tip to 193 mm at the top and contained absolute, strain gauge, pressure transducers, with a 0.5 MPa range, located at 7 levels along its length with the lowest sensor positioned about 13 m below the mudline. Very coarse filters made from perforated nylon discs were used as pressure ports on the probe. The piezometer was jacked into the sandy and clayey seabed in a mean water depth of about 8 m with a tidal range of about 5 m in Oct/Nov 1981, and was removed in June 1983. Data were collected at different times throughout the deployment and were checked at least once per week. Dynamic pore pressures from both waves and tides were measured at depths up to 13 m below the seabed, although it was considered difficult to draw any firm conclusions from the data (De Wolf *et al.*, 1983).

NORDA DEEP OCEAN PIEZOMETER

A piezometer probe capable of working in the deep ocean (up to 6000 m) has been developed at the Naval Ocean Research and Development Activity (NORDA) (Bennett *et al.*, 1985). The development was an integral part of the *In Situ* Heat Transfer Experiment (ISHTE) which was conducted in sup-

port of the US sub-seabed disposal programme examining the feasibility of emplacing high-level nuclear waste in the deep seabed. The titanium probe tip is only 8 mm in diameter, has a sharp cone (angle approximately 5°), and contains a porous stone as a pressure port that leads via a pipe to a variable reluctance differential pressure transducer (range = +/−69.8 kPa, precision less than 340 Pa). The piezometer was initially tested using the submersible ALVIN in 1450 m of water off the US Atlantic continental slope (Lambert, 1982). However, the pore pressure sensor (a differential semi-conductor type was used at that time) failed completely soon after insertion. The probes were subsequently used during a laboratory experiment designed to simulate the environmental conditions for the ISHTE experiment. The probes were inserted into a tank (approx. 1 m^3) filled with a fine-grained sediment at a hydrostatic pressure of 55 MPa and at 4°C in November 1981, to depths of 16.9 cm and 26.4 cm below the mudline. Maximum insertion pressures of 6.6 kPa and 12.9 kPa were recorded before the pressure smoothly decayed to near zero values in times of 20 min. and 61.5 min. respectively.

Abbott *et al.* (1984) have measured pore pressures in deep-water sediments using differential pressure transducers mounted on a heat flow probe in the Gautemala Basin. They postulated a maximum downward pore water advection velocity of between 30 and 300 mm/year in those areas with thinest sediment cover. However, their residual pore pressure values were based on extrapolating the decay curve of the transient pressure pulse caused by insertion, using an asymptotic computer-generated model with less than 10 min. of data. To eliminate the uncertainties of theoretical extrapolation, much longer-term deployments are needed in order that the insertion pressure decays more completely before the residual pore pressure is measured.

PUPPI

An autonomous Pop-Up-Pore-Pressure-Instrument (PUPPI) has been designed and constructed to accurately measure *in situ* pore pressure gradients in deep-sea sediments (Schultheiss *et al.*, 1985; McPhail and Schultheiss, 1986). The PUPPI measures differential pore pressures at two ports with a resolution of +/−15 Pa on a lance up to 6 m long. The PUPPI is a free-fall instrument that can be ballasted to

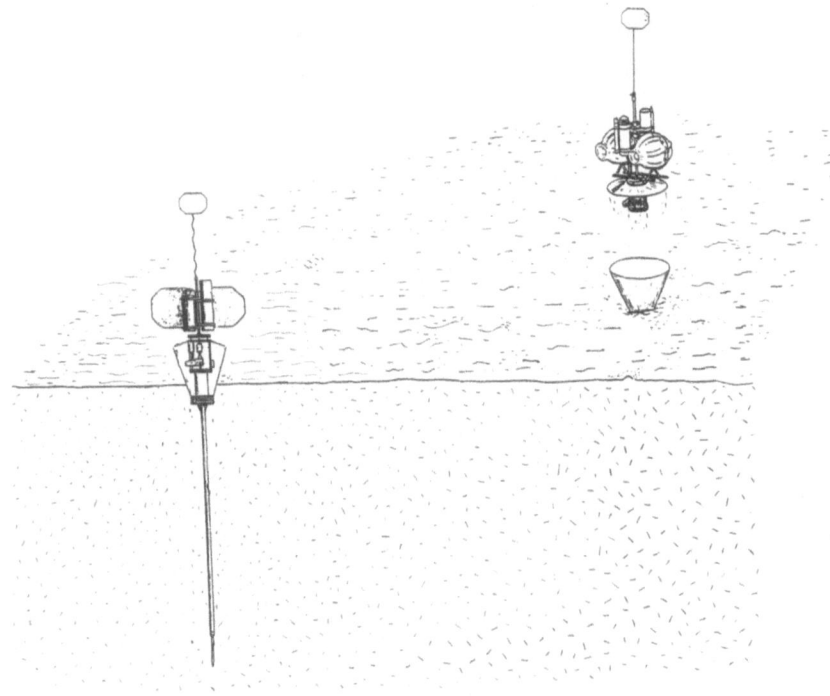

Fig. 5. The Pop-Up-Pore-Pressure-Instrument (PUPPI) embedded in the sediments, after free-falling through the water column, and after the upper buoyant section has been acoustically released.

penetrate a range of sediment types in water depths of up to 6000 m. Pore pressures are measured relative to hydrostatic at the ports on the lance by using differential pressure transducers connected between the ports and open sea water. A programmable solid-state logger is used to record the pressure data at a maximum rate of 0.5 samples/sec. A vertically mounted accelerometer enables the penetration event to be examined and allows the depth of penetration to be calculated. Recovery is accomplished using an acoustic command which activates a release mechanism that simultaneously cuts the pipe connecting the pressure port to the transducer. The lance, ballast weight and cone assembly are left on the sea floor while the buoyant instrument package ascends to the surface for recovery (Fig. 5).

To date, 40 deployments have been made using PUPPI systems in water depths between 3500 m and 5500 m. Most of the deployments have been in the Madeira Abyssal Plain (Schultheiss and Noel, 1987), where differential pore pressures were found to be generally absent. However in the Mariana back arc basin, significant pore pressure gradients were measured, indicating local upward flow of pore water (Leinen *et al.*, in prep.). Data have also been col-

lected from the Barbados Accretionary Complex indicating significant pervasive flow of pore water.

A generalized diagram of the pore pressure data obtained from the PUPPI is shown in Fig. 6. It illustrates the pertinent features of the pore pressure record that can be used to infer some *in situ* properties of the surrounding sediment. These features are discussed below.

(a) The maximum insertion pressure, U_{max}, generated at the probe surface has been modelled by

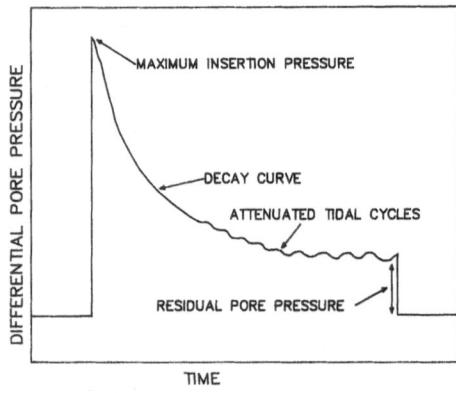

Fig. 6. Generalized pore pressure record from the PUPPI showing the four main pressure features that are used to calculate some of the *in situ* properties of sediments.

Randolf *et al.* (1979). Esrig *et al.* (1977) suggested that some simple generalizations could be made for different soil types to predict the *in situ* undrained shear strength, C_u, from U_{imax}. For lean inorganic soils of moderate-to-high sensitivity (typical of many deep sea clays) this resulted in the simple expression:

$$C_u = U_{imax}/6.$$

(b) The dissipation of pore pressure (the decay curve) can be modelled as radial consolidation around a cylindrical probe. Bennett *et al.* (1985) uses the solutions of Soderberg (1962) and Randolf *et al.* (1979) to predict the coefficient of horizontal consolidation, C_h, from the time taken for 50% of U_{imax} to dissipate (t_{50}) using the expression:

$$C_h = r^2 T_{r50}/t_{50},$$

where r is the radius of the probe and T_{r50} is a dimensionless time factor which is approximately equal to one for this case. Having obtained C_h, the permeability, k, can be determined using appropriate values of the constrained modulus, D, and the unit weight of water γ_w from the expression:

$$k = C_h \gamma_w /D.$$

(c) The attenuation of dynamic hydrostatic pressure in the sediments caused by tidal cycles can be modelled as standing waves, as discussed previously. This modelling enables *in situ* values of permeability and shear modulus to be predicted.

(d) The residual differential pore pressure that exists after the complete decay of the insertion pressure is the ambient *in situ* pressure. This affects the *in situ* state of effective stress and will cause water to flow in the permeable sediments. Consequently it can be used to determine the natural pervasive flux rates of pore waters.

Pore Pressure Measurements—Practical Problems

There are many potential problems associated with the accurate measurement of pore pressures in sediments such as time-lag, resolution, accuracy and calibration all of which can change with time, temperature and pressure. A discussion of most of these problems is beyond the scope of this chapter, but the reader is referred to a useful review of pore pressure

measurement devices for use in terrestrial soils by Hanna (1985). One of the major problems of calibration has been overcome on the PUPPI by remotely cutting the pipe leading to the pressure port, immediately before recovery. In this way both sides of the differential transducer are open to the sea water pressure, and consequently a very accurate *in situ* zero calibration is obtained.

For dynamic measurements, the time-lag associated with the pressure measurement system is of critical importance. A perfect pressure transducer would instantaneously and accurately measure a change in pore pressure within the sediments no matter how sudden. In practice, a finite flow of water must move into or out of the transducer before equilibrium is reached and the transducer registers the true change in pressure. Hence there will always be a time-lag between a pore pressure change in the sediments and the transducer reaching its equilibrium state. The time-lag of a piezometer system in a sediment depends on several factors, including the permeability of the surrounding sediments.

Hvorslev (1951) has given a solution to the time-lag for any shape of piezometer in an incompressible soil. However, the assumption of an incompressible soil can lead to appreciable errors as recognized by Penman (1960). A solution for the time-lag of spherical tip piezometers in compressible soils has been derived by Gibson (1963), using the coefficient of consolidation of the soil.

In practice, the most important factor to consider, in order to minimize the time-lag of a piezometer, is to ensure that it is fully saturated. Small amounts of air, or gas, in any part of the system (pressure port, pipes or sensor body) can significantly increase the overall volume compliance, and consequently the amount of water flow into and out of the system. With differential pressure transducers the effects can be even more dramatic. If the overall compliance (diaphragm, pipework and trapped air) is not the same on both sides of the sensing element, then the pressure readings will be seriously in error both in amplitude and phase.

MEASUREMENTS IN GASSY SEDIMENTS

It is obvious from the above comments that for accurate dynamic measurements of pore pressure in sediments the pressure measurement system must be saturated. When working in deep water achieving saturation is not difficult, as any small amounts of gas left in the system will be dramatically reduced in volume and will finally be forced into solution as the instrument descends in the water. With good design and care, full saturation of a measurement system can be achieved even in shallow water. Problems really begin to occur when the sediments themselves are not saturated.

The pore pressure measurements made during the SEASWAB experiments clearly demonstrate the problems of intepreting pressure measurements in gassy sediments. As previously discussed, even the use of HAE pressure ports did not prevent gas diffusing through into the pipework and sensor.

In fine-grained sediments that contain free gas it can reasonably be assumed that the pore water is saturated with gas, that the gas bubbles are probably much larger than the sediment pore sizes and that these bubbles are relatively stable within the sediment framework. Under these conditions the free gas cannot easily migrate from the sediment through the porous pressure port and into the pipework and sensor directly. However, the gas concentration gradient will cause gas in solution to diffuse through the pressure port and into the sensor. The fluid inside the sensor will eventually become saturated in gas and will consequently be in chemical equilibrium with the pore waters in the sediment. If this is the case, then under conditions of static ambient pore pressures, the pressure registered by the sensor would accurately reflect the pore pressure in the sediments. If, however, the gas in the sensor were to come out of solution, then serious errors could result if the gas replaces the water over a significant length of the pipe. If the pressure in the sediments is varying dynamically, as in the case where waves and tides have an effect, then the errors caused by even small amounts of gas could be very large indeed. It may be the dynamic effects of pressure and/or temperature that enhance the diffusive flux of gas into the sensor and cause the gas to come out of solution after it has reached saturation (T. R. S Wilson, pers. comm.). Therefore, any pore pressure measurements made in gassy sediments will probably be very difficult to interpret accurately after a couple of days. The author is unaware of any satisfactory simple solution to this problem.

Summary and Conclusions

Pore pressures are not only a fundamentally important parameter for understanding the behaviour of soils, they are also extremely sensitive to imposed stress, especially in soils with low permeabilities. *In situ* measurements of both static and dynamic pore pressures can provide crucial information concerning the rate at which pore fluids move, and important data on the *in situ* properties of sediments. This type of information has significant applications in both geological and engineering studies of the sea floor. Many of the perceived problems in the field of marine hydrogeology can, at least partially, be addressed from measurements of pore pressure at relatively shallow depths within the sediments. These range from investigating major fluid flow processes within the ocean crust to studying the sedimentation and consolidation phenomena of sediments in the relatively static oceanic basins. For engineering projects that require detailed site characterization, *in situ* effective stresses must be calculated; hence, pore pressures must be known. They must also be known if the dynamic effects of storms and tides on slope stability are to be assessed more accurately. Hydrodynamic models are available that enable important parameters such as permeability and shear modulus to be calculated from the dynamic pore pressure response of the sediments to tidal cycles.

The technology for very accurately measuring differential pore pressures in the upper few metres of soft sediments now exists even for use in full oceanic water-depths. Simple techniques for measurements at greater sub-bottom depths and in harder substrates are not available. One of the largest outstanding problems in soft sediments lies with measuring pore pressures in gassy sediments over any significant period of time.

Much more data are required from different sedimentary and tectonic environments to validate the models, to appreciate further the nature of some of the outstanding measurement problems and to enhance our understanding of the varied, and probably complex, hydrogeological processes occurring in the marine environment.

Acknowledgements

I would like to thank my colleagues at the Institute of Oceanographic Sciences for their encouragement and help with the development and use of the PUPPI. In particular, I wish to thank Steve McPhail, Dave Gunn, Tim Francis, Mike Hurley and Richard Babb for their particular skills. Thanks are also given to Ernie Hailwood who carefully reviewed the original manuscript.

References

Abbott, D., Menke, W., Hobart, M., Anderson, R. N., and Embley, R. W., 1984, Correlated Sediment Thickness, Temperature Gradient and Excess Pore Pressure in a Marine Fault Block Basin. *Geophys. Res. Lett.* **11**(5), 485–488.

Baligh, M. M., 1986, Undrained Penetration, II: Pore Pressures, *Geotechnique* **36**(4) 487–501.

Becker, K. and Von Herzen, R. P., 1983, Heat Transfer through the Sediments of the Mounds Hydrothermal Area, Galapagos Spreading Center at 86°W, *J. Geophys. Res.* **88**, 995–1008.

Bennett, R. H., Bryant, W. R., Dunlap, W. A., and Keller, G. H., 1976, Initial Results and Progress of the Mississippi Delta Sediment Pore Water Pressure Experiment, *Marine Geotechnology* **1**(4), 327–335.

Bennett, R. H., 1977, Pore Water Pressure Measurements: Mississippi Delta Submarine Sediments, *Marine Geotechnology* **2**, 177–189.

Bennett, R. H. and Faris, J., Ambient and Dynamic Pore Pressures in Fine-grained Submarine Sediments: Mississippi Delta, *Applied Ocean Research* **1**(3) 115–123.

Bennett, R. H., Burns, J. T., Clarke, T. L., Faris, J. R., Forde, E. B., and Richards, A. F., 1982, Piezometer Probes for Assessing Effective Stress and Stability in Submarine Sediments, in *Marine Slides and other Mass Movements* (Saxov, S. and Nieuwenhuis, J. K., eds), Plenum Press, London, pp. 129–161.

Bennett, R. H., Huon, L., Valent, P. J., Lipkin, J., and Esrig, M. I., 1985, *In Situ* Undrained Shear Strengths and Permeabilities Derived from Piezometer Measurements, in Chaney, R. C. and Demars, K. R. (eds), *Strength Testing of Marine Sediments: Laboratory and In Situ Measurements* Special Technical Publication 883, American Society for Testing and Materials, Philadelphia, pp. 88–100.

Bennett, R. H., Burns, J. T., Nastav, F. L., Lipkin, J., and Percival, C. M., 1985, Deep-Ocean Piezometer Probe Technology for Geotechnical Investigations, *IEEE Journal of Oceanic Engineering* **10**(1), 17–22.

Carpentier, R. and Verdonk, W., 1986, Special Pore Water Pressure Measuring System Installed in the Seabed for the Construction of the New Outer Harbour at Zeebrugge, Belgium, *International Conference on Measuring Techniques*, London, paper C3, pp. 121–135.

De Wolf, P., Carpentier, R., Verdonk, W., Boullart, L., De Rouk, J., de Saint Aubain, T., and De Voghel, J., 1983, In-situ Pore Water Pressure Measurements for the Construction of the Breakwaters of the New Outer Harbour at Zeebrugge, *Proceedings 8th International Harbour Conference*, Antwerp, pp. 1203–1215.

Dunlap, W. A., Bryant, W. R., Bennett, R., and Richards, A. F., 1978, Pore Pressure Measurements in Underconsolidated Sediments. 10th Offshore Technology Conference Proceedings, **2**, paper 3168, pp. 1049–1058.

Esrig, M. I., Kirby R. C., and Bea, R. G., 1977, Initial Development of a General Effective Stress Method for the Prediction of

Axial Capacity for Driven Piles in Clay, 9th Offshore Technology Conference Proceedings, paper 2943, pp. 495–501.

Francis, T. J. G., 1984, A Review of IOS Research into the Feasibility of High-Level Radioactive Waste Disposal in the Oceans, *The Science of the Total Environment* **35**, 301–323.

Garrison, L. E., 1977, The SEASWAB Experiment, *Marine Geotechnology* **2**, 117–122.

Gibson, R. E., 1963, An Analysis of the System Flexibility and its Effect on the Time Lag in Pore Water Pressure Measurements, *Geotechnique* **13**(1), 1–11.

Hamilton, E. L., 1964, Consolidation Characteristics and Related Properties of Sediments from Experimental Mohole (Guadalupe Site), *J. Geophys. Res.* **69**(20), 4257–4269.

Hanna, T. H., 1985, Pore Water Pressure Measurement Devices, in Hanna T. H. (ed.), *Field Instrumentation in Geotechnical Engineering*, Clausthal-Zellerfeld Trans. Tech. Publications, 843 pp, *Series on Rock and Soil Mechanics* **10**, 121–204.

Hirst, T. J. and Richards, A. F., 1976, Excess Pore Pressure in Mississippi Delta Front Sediments: Initial Report, *Marine Geotechnology* **1**(4), 337–344.

Hirst, T. J. and Richards, A. F., 1977, *In situ* Pore Pressure Measurements in Mississippi Delta Front Sediments, *Marine Geotechnology* **2**, 191–204.

Hottman, W. E., Suhayda, J. N., and Garrison, L. E., 1978, SEASWAB II (Shallow Experiment to Assess Storm Wave Effects on the Bottom), *Offshore Technology Conference Proceedings*, **2**, paper 3169, 1059–1066.

Hurley, M. T., 1989, *Biot's Theory of Poroelasticity Applied to Geoacoustic and Hydrodynamic Models of the Sea-bed*, Unpublished PhD Thesis, University of Wales.

Hvorslev, M. J., 1951, Time Lag and Soil Permeability in Ground Water Observations. *Bulletin No. 36*, Waterways Experimental Station, Corps of Engineers, U.S. Army.

Lai, J. Y., Richards, A. F., and Keller, G. H., 1968, In-Place Measurement of Excess Pore Water Pressure of Gulf of Main Clays (Abstract), *American Geophysical Union Transactions* **49**, 221.

Lambert, D. N., 1982, Submersible Mounted *in situ* Geotechnical Instrumentation. *Geo. Marine Letters* **2**, 209–214.

Leinen, M., Schultheiss, P. J., Dadey, K., Becker, K., and McDuff, R., 1990, *Ridge Flank Hydrothermal Activity near the Mariana Trough Spreading Center* (in prep.).

McPhail S. D. and Schultheiss, P. J., 1986, PUPPI—A Free-Fall Seabed Piezometer for Geotechnical Studies, *Advances in Underwater Technology, Ocean Science and Offshore Engineering, Vol. 6: Oceanology*, Graham and Trotman Ltd., pp. 249–258.

Noorany, I. and Gizienski, S. F., 1970, Engineering Properties of Submarine Soils—A State-of-the-Art Review, *Proc. American Society of Civil Engineers* **96**, (SM5) 1735–1762.

Penman, A. D. M., 1960, A Study of the Response Time of Various Types of Piezometer, Proceedings, Conference on Pore Pressure and Suction in Soils, *Institution of Civil Engineers*, London, pp. 53–58.

Prindle, R. W. and Lopez, A. A., 1983, Pore Pressures in Marine Sediments—1981 Test of the Geotechnically Instrumented Seafloor Probe (GISP), *Offshore Technology Conference Proceedings*, Vol. 1 (4463), 173–180.

Prior, D. B. and Coleman, J. M., 1984, Submarine Slope Instability, in Brunsden, D. and Prior, D. B., (eds.) *Slope Instability*, John Wiley & Sons Ltd., pp. 419–455.

Randolf, M. F., Carter, J. P. and Wroth, C. P., 1979, Driven Piles in Clay—The Effects of Installation and Subsequent Consolidation, *Geotechnique* **29**(4), 361–393.

Reece, E. W., Ryerson, D. E., Kestly, J. D., and McNeill, R. L., 1979, The Development of *in situ* Marine Seismic and Geotechnical Instrumentation Systems. POAC 79, in *Proceedings of 5th International Conference on Port and Ocean Engineering under Arctic Conditions*, **1**, The Norwegian Institute of Technology, Trondheim, pp. 331–334.

Richards, A. F., 1984, Modelling and the Consolidation of Marine Soils, in Denness, B. (ed.), *Seabed Mechanics*, Graham and Trotman, pp. 3–8.

Richards, A. F. and Hamilton, E. L., 1967, Investigations of Deep Sea Sediment Cores, III. Consolidation, in Richards, A. F., (ed.), *Marine Geotechnique*, University of Illinois Press, Urbana, pp. 93–117.

Richards, A. F., Oien, K., Keller, G. H., and Lai, J., 1975, Differential Piezometer Probe for *in situ* Measurement of Seafloor Pore Pressure, *Geotechnique* **25**, 229–238.

Schultheiss, P. J., McPhail, S. D., Packwood, A. R., and Hart, B., 1985, An Instrument to Measure Differential Pore Pressures in Deep Ocean Sediments: Pop-up-Pore-Pressure-Instrument, (PUPPI). IOS Report No. 202, 57 pp.

Schultheiss P. J. and McPhail, S. D., 1986, Direct Indication of Pore-Water Advection from Pore Pressure Measurements, in Madeira Abyssal Plain Sediments, *Nature* **320**(6060), 348–350.

Schultheiss, P. J. and Noel, M., 1987, Evidence for Pore Water Advection in the Madeira Abyssal Plain from Sediment Pore-Pressure and Temperature Measurements, in P. P. E. Weaver and J. Thomson, (eds.), *Geology and Geochemistry of Abyssal Plains*, *Geological Society Special Publication* **31**, 113–129.

Sills, G. C. and Nageswaran, S., 1984, Compressibility of Gassy Soil, *Oceanology International*, Brighton, UK, Paper OI2.6/1., 18 pp.

Silva, A. J. and Jordan, S. A., Consolidation Properties and Stress History of Some Deep Sea Sediments, in Denness, B., (ed.), *Seabed Mechanics*, Graham and Trotman, pp. 25–39.

Silva, G. C. and Nageswaran, S., 1984, Compressibility of Gassy Soil, *Oceanology International*, Brighton, UK, paper OI2.6/1., 18 pp.

Silva, A. J. and Jordan, S. A., 1984, Consolidation Properties and Stress History of some Deep Sea Sediments, in Denness, B., (ed.), *Seabed Mechanics*, Graham and Trotman, pp. 25–39.

Silva, A. J. and Mairs, H. L., 1987, Stress History and Geotechnical Properties of Nares and Madeira Abyssal Plain Sediments, Marine Geomechanics Laboratory, Dept. of Ocean Engineering, University of Rhode Island Technical Report, 161 pp.

Soderberg, L. O., 1962, Consolidation Theory Applied to Foundation Pile Time Effects, *Geotechnique* **12**, 217–225.

Terzaghi, K., 1943, *Theoretical Soil Mechanics*, John Wiley & Sons, New York.

Vaughn, P. R., 1973, The Measurement of Pore Pressures with Piezometers, in *Field Instrumentation in Geotechnical Engineering*, John Wiley, NY, pp. 411–422.

Westbrook, G. K. and Smith, M. J., 1983, Long Decollements and Mud Volcanoes: Evidence from the Barbados Ridge Complex for the Role of High Pore Water Pressures in the Development of an Accretionary Complex, *Geology* **11**, 279–283.

Working Group 3, 1987, Fluid Circulation in the Crust and Global Geochemical Budget, in *Report of the Second Conference on Scientific Ocean Drilling* (*COSOD II*), Strasboug, France, European Science Foundation, pp. 67–86.

List of Contributors

J. GRANT*, Krupp Atlas Elektronik UK Ltd., Hartcran House, Delta Gain, Carpenders Park, Watford, WD1 5EZ, UK

E. A. HAILWOOD*, Oceanography Department, University of Southampton, Southampton, SO9 5NH, UK

Q. J. HUGGETT*, Institute of Oceanographic Sciences Deacon Laboratory, Brook Road, Wormley, Godalming, Surrey, GU8 5UB, UK

R. B. KIDD*, Geology Department, University College of Wales, P.O. Box 68, Cardiff CF1 3XA, UK

T. P. LeBAS, Institute of Oceanographic Sciences Deacon Laboratory, Brook Road, Wormley, Godalming, Surrey, GU8 5UB, UK

M. R. G. MacCORMACK, Department of Earth Sciences, University of Cambridge, Bullard Laboratories, Madingley Road, Cambridge, CB3 0EZ, UK

N. C. MITCHELL, Department of Earth Sciences, University of Oxford, Parks Road, Oxford OX1 3PR, UK

T. R. E. OWEN, Department of Earth Sciences, University of Cambridge, Bullard Laboratories, Madingley Road, Cambridge, CB3 0EZ, UK

W. R. PARKER*, Blackdown Consultants Ltd., 3, Curdleigh Lane, Blagdon Hill, Taunton, Somerset, TA3 7SH, UK

L. M. PARSON, Institute of Oceanographic Sciences Deacon Laboratory, Brook Road, Wormley, Godalming, Surrey, GU8 5UB, UK

P. D. PATEL, Department of Earth Sciences, University of Cambridge, Bullard Laboratories, Madingley Road, Cambridge, CB3 0EZ, UK

Ph. PATRIAT, Institut de Physique du Globe de Paris, Tour 14–15, 4 Place Jussieu, 75230, Paris, France

A. T. S. RAMSAY, Geology Department, University College of Wales, P.O. Box 68, Cardiff, CF1 3XA, UK

J. M. REYNOLDS*, Department of Geological Sciences, Polytechnic South West, Drake Circus, Plymouth, PL4 8AA, UK

C. M. R. ROBERTS*, Department of Geological Sciences, University of Durham, South Road, Durham, DH1 3LE, UK

K. ROBERTSON*, NERC Scientific Services, Research Vessel Services, No 1 Dock, Barry, South Glamorgan CF6 6UZ

R. SCHREIBER, Krupp Atlas Elektronic GmbH, Sebaldsbrücker Heerstr. 235, D2800, Bremen 44, FRG

P. J. SCHULTHEISS*, Geotek, Fern Cottage, Marley Lane, Haslemere, Surrey GU27 3RF, UK

R. C. SEARLE*, Department of Geological Sciences, University of Durham, South Road, Durham, DH1 3LE, UK

G. C. SILLS, Department of Engineering Science, University of Oxford, Parks Road, Oxford, OX1 3PJ, UK

M. C. SINHA*, Department of Earth Sciences, University of Cambridge, Bullard Laboratories, Madingley Road, Cambridge, CB3 0EZ, UK

M. L. SOMERS, Institute of Oceanographic Sciences Deacon Laboratory, Brook Road, Wormley, Godalming, Surrey, GU8 5UB, UK

M. STORMS*, Ocean Drilling Program, Texas A&M University Research Park, 1000 Discovery Drive, College Station, Texas 77840, USA

M. J. UNSWORTH, Department of Earth Sciences, University of Cambridge, Bullard Laboratories, Madingley Road, Cambridge, CB3 0EZ, UK

P. P. WEAVER*, Institute of Oceanographic Sciences Deacon Laboratory, Brook Road, Wormley, Godalming, Surrey, GU8 5UB, UK

J. WHITE*, Department of Civil Engineering, University of Southampton, SO9 5NH, UK

* = first author.